ALGEBRA AND TRIGONOMETRY REFRESHER
FOR CALCULUS STUDENTS

KEY TO CALCULUS TOPICS

Your algebra and trigonometry review can be spread over the entire calculus course. Before you do your calculus assignment, take a few minutes to work through the appropriate sections recommended below. The examples and exercises in the REFRESHER feature the skills needed for the day's calculus lesson.

Calculus Topic	REFRESHER Sections
Absolute Value, Inequalities	9, 19
Lines and Graphs	7, 8
Circles, Ellipses, Parabolas, Hyperbolas	25, 26
Functions	15, 23
Limits	12, 13
Derivatives (Using the Definition)	15, 16
Derivatives (The Power Rule)	10, 14
Derivatives (The Chain Rule)	15
Derivatives (Simplification)	12, 13, 16
Implicit Differentiation	16
Curve Sketching Using the Derivative	17, 18, 19
Word Problems	21, 22, 23
Length, Area, Volume by Integration	20, 21
Growth and Decay Problems	27
Logarithms, Logarithmic Differentiation	27, 28
Trigonometric Functions	29, 30, 31, 32
Inverse Trigonometric Functions	35
Partial Fractions	20
Polar Coordinates	29, 31
Integration by Trigonometric Substitution	34
Calculus Applications Using Trigonometric Functions	33
Integration by Inverse Trigonometric Substitution	35
Use of Integral Tables	25
Linear Differential Equations	18, 36

We're supposed to know this?

ALGEBRA
AND TRIGONOMETRY
REFRESHER
FOR CALCULUS STUDENTS

Loren C. Larson

St. Olaf College

W. H. Freeman and Company
San Francisco

Library of Congress Cataloging in Publication Data

Larson, Loren C 1937-
   Algebra and trigonometry refresher for calculus
students.

   (A Series of books in the mathematical sciences)
   Bibliography: p.
   Includes index.
   1. Algebra. 2. Trigonometry. I. Title.
QA154.2.L37   512'.13   79-20633
ISBN 0-7167-1110-9

Printed in the United States of America

   2 3 4 5 6 7 8 9 0

To   JEN JENKINS

        Professor Emeritus

        Bethany College, Lindsborg, Kansas

        who made this subject come alive

        for hundreds of students

A SERIES OF BOOKS IN THE MATHEMATICAL SCIENCES

Victor Klee, Editor

PREFACE

Because the solution of calculus problems requires the use of algebra or trigonometry, success in calculus is directly related to mastery of these two subjects. Students with strong backgrounds in algebra and trigonometry are free to concentrate on the calculus. Students with weaker backgrounds, however, will be unable to solve problems in calculus, even when they may have mastered calculus concepts. Regardless of how well they have learned the prerequisite courses, many students may have forgotten the techniques and formulas included in these courses.

These notes are a reminder of the central ideas, concepts, formulas, and problem-solving techniques from algebra and trigonometry essential for progress in calculus. The text is divided into thirty-six sections arranged in four parts. Each section consists of three or four appropriate examples with just enough discussion to jog the memory, followed by a few appropriate exercises (and answers). Throughout, the exercises and examples reflect the algebraic and trigonometric problems encountered in calculus. Some problems are more difficult than those common in algebra and trigonometry courses. However, calculus students must learn to handle this material with confidence if they are to be successful. This REFRESHER not only allows the student to review and learn the prerequisites for calculus and its use, but it also allows the instructor to define those prerequisites clearly and explicitly by referring the student to the appropriate sections of this book.

Although the choice and organization of topics is dictated by the demands of the calculus course, this book may be used for other courses as well. Because of its wide coverage and its organization into brief, independent sections, and its practical, problem-solving approach, the book may serve various purposes. For example, it is suitable as a supplement for all beginning science courses (including the behavioral sciences). More specifically, science courses for nonscience majors require familiarity with the topics of Sections 1 through 10, 20, 21, 22, and 24, while the requirements for physics and chemistry are the same as those given here for calculus. Students preparing for entrances exams for college or for graduate or professional schools will find the notes useful as a study guide.

This writing project was supported in part by the Northwest Area Foundation. I wish to express my gratitude to the following people, who offered valuable comments and suggestions: Victor Klee from the University of Washington and Peter Renz and Andrew Kudlacik from W. H. Freeman and Company. From St. Olaf College, I wish to thank Lorraine Keller, Director of Freshman Mathematics; Judy Cederberg, Instructor of Mathematics; Martha Wallace, Analytical Skills Coordinator; Frances Robinson, who also did the typing; and Al Magnuson, who produced the diagrams.

Loren C. Larson
February 1979

TO THE TEACHER

Even in the first few days of class it is easy to identify students who are almost certain to have difficulty with calculus. Simply ask your students to do a few routine algebra-trigonometry problems, perhaps at the blackboard. Look for the familiar signals: inaccurate and unreliable calculations; slow and inefficient manipulation; sloppy, unorganized, hard-to-read work. Students with these symptoms are headed for trouble. Short of sending them back to a precalculus course, what can you do to help prepare the student for the difficulties that lie ahead? In many students, the weaknesses are general and cannot be remedied by suggesting, for instance, more work on quadratic equations. General advice, such as to "review your algebra and trigonometry, a few problems each day," is too vague and will not be followed. The student's long-standing mathematical deficiencies cannot be dealt with in a single help session or even in a short series of such sessions. These notes provide a specific, concrete plan of action for spreading the necessary review over the entire calculus course. They are self-diagnosing, and will be on hand when the student needs them for learning, review, and reference. Here are some of the ways an instructor can make use of the notes.

- Make available diagnostic tests such as those at the beginning of each of the four parts of these notes. Students who need review and practice will find both in these notes.

- During the course, hold three or four special help sessions based on these notes to review the prerequisite mathematics.

- Use these examples and exercises in the classroom to highlight the constituent algebraic and trigonometric ideas required for the day's lesson.

- Make daily assignments from the notes (coordinated with the
  calculus assignment) and hold students accountable for the
  mastery of this content. Assignments from the notes are
  easy to make: simply assign *all* the exercises in the section. The sections and exercise sets are brief and to the
  point.

At St. Olaf College, incoming freshmen receive a description of the beginning mathematics curriculum in the early summer before they enroll. Prospec-
tive calculus students are informed of our expectations and are encouraged to
use the REFRESHER, which is available through the college bookstore.

TO THE STUDENT

The algebra and trigonometry contained in these notes are essential for calculus. You can be assured that this material will be used in the course, and your efforts at mastery will pay dividends in understanding.

In using these notes, first concentrate on the examples. Even though this is a review, don't expect to be able to produce an immediate solution to each example. Relax and work your way carefully through the solution, proceeding at your own pace. Make certain you follow the details of the solution, and concentrate on understanding the overall method. If you have difficulty with either the details or the approach, relax and try again. If, after a reasonable effort, you are still perplexed about some point, make a note of it, then read on and come back to it later. Proceed in this way through the entire section.

After reading the section and the examples, you should work the exercises. These check your understanding of the examples, develop your skills, and instill confidence in your ability to handle the ideas. *If you want to improve*, you must write out the solutions to the exercises in detail, so that they are neat and well organized. It is not enough to murmur hastily, "Of course, quite simple," and go on to the next exercise. On tests it is not always "quite simple," and your instructors will require clear, coherent solutions. Practice is the only way to gain facility in algebraic and trigonometric manipulation.

If you can't work a particular exercise, don't give up. Reread the examples for clues. If the solution still doesn't come to you, you will at least have specific questions to bring to your instructor, tutor, or fellow student. Answering such specific questions is a pleasure and can be done quickly and effectively, saving you and your instructor time and effort.

If there are gaps in your background, you may find some material in the REFRESHER difficult to master. In some cases, consult the references at the end of the book (or any standard algebra, trigonometry, or precalculus text) for further explanation.

Mastery will come gradually. You may not notice improvement every day. However, if you work at it regularly--for example, as a part of each homework assignment---you will soon gain the necessary mathematical skills. Lucy Sells has aptly called the lack of such skills the "critical filter that bars women [and men] from scientific and technical occupations."*

*Lucy Sells, "High School Mathematics as the Critical Filter in the Job Market," in *Developing Opportunities for Minorities in Graduate Education*, Proceedings of the Conference on Minority Graduate Education, University of California, Berkeley, May 11-12, 1973, pp. 37-39.

CONTENTS

(The objectives for each section are illustrated by

the problems.   See pages 185-187 for answers.)

PART I: Preliminaries

Simplify: (i) $3a - 4b - 2 - 2a - 5b - 7$

(ii) $2\sqrt{3} + 5\sqrt{5} - 2\sqrt{5} - 7\sqrt{3}$

Simplify: (i) $2 - (4(3 - 5) + 3)$

(ii) $2(7a - 5b) - 3(3a + 4b)$

Combine and simplify: (i) $(\frac{2}{3} - \frac{1}{2}) - \frac{5}{6}$

(ii) $(\frac{2}{3a} - \frac{1}{2b}) - \frac{5}{6ab}$

Perform the operations and simplify: (i) $\frac{3/4 + 1/3}{2/3 - (-1/2)}$

(ii) $(\frac{2}{b} - \frac{3}{a}) \div (\frac{b}{2} - \frac{a}{3})$

Solve for x: (i) $30 = \frac{5}{9}(x - 32)$

(ii) $\frac{4x - 23}{6} + \frac{1}{3} = \frac{5x}{4}$

PART II: Algebra for Derivatives

PART III: Algebra for Applications

Solve for x: (i) $x(2x + 1) = 6$

(ii) $x = \dfrac{2x + 1}{x + 1}$

(i) Find the solutions to $x^3 - x^2 - 8x + 12 = 0$, given that

x = 2 is a solution.

(ii) Solve for x: $2(x + 1)(2x + 3)^3 + 6(x + 1)^2(2x + 3)^2 = 0$.

Solve the inequalities: (i) $(x - 3)(x + 2) \leq 0$

(ii) $\dfrac{3}{x - 1} \leq 1$

Solve: (i) $2x + 3y = 4$ (ii) $y = 4 - x^2$

$3x - 4y = 5$ $y = -3x$

(i) If the circumference of a circle is 10π, what is its area?

(ii) In the figure at the right, find

a relationship between x and y.

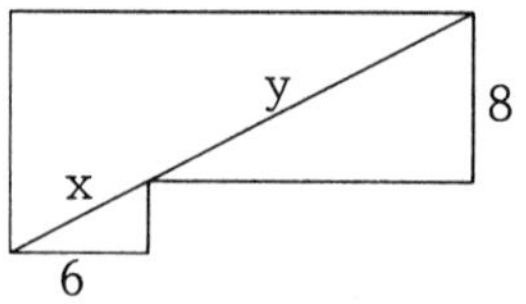

(i) A racing track consists of a rectangle of length ℓ and width

w with a semicircle on each end. Write a formula for the

perimeter of the track. Suppose the perimeter of the track

(the length of the curbing) measures exactly 440 yards (1/4 mile). How long is one lap for a runner in the inside lane who runs 12 inches from the curb?

(ii) The strength S of a rectangular beam is proportional to the width w and square of the depth d. Express this algebraically. Compare the strength of a 4 x 6 beam when it is placed on its edge with its strength when it is turned on its side.

(i) Two posts are placed 25 feet apart. One post is 8 feet high and the other is 12 feet high. A guy wire is attached to the top of each post and to a stake at ground level between them. Express the length of the wire as a function of the distance between the 8 foot post and the stake.

(ii) A conical container, with its vertex pointed down, is 6 inches deep and 4 inches across. A small amount of salt water is poured into this container. Express the surface area of the salt water as a function of the depth.

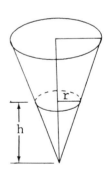

(i) How many cubic meters are in a tray whose measurements are 4 feet by 3 feet by 6 inches (1 inch = 2.54 centimeters)?

(ii) Which is the faster average speed: 100 meters in 10 seconds or 100 yards in 9.1 seconds?

PART IV: Logarithms, Trigonometry

    (i) Write $\dfrac{3 + i}{1 + 2i}$ in the form a + bi, a and b real.

    (ii) If z is a complex number of magnitude 2 and argument 40°,

       what is the magnitude and argument of z^3?

This self-diagnostic table of contents is from ALGEBRA
AND TRIGONOMETRY REFRESHER FOR CALCULUS STUDENTS by
Loren C. Larson. Copies of this book may be purchased
from your bookstore or from the publisher:

    W. H. Freeman and Company

    660 Market Street

    San Francisco, California 94104

PART I: Preliminaries

Before you begin the study of calculus, you should feel comfortable with the ideas presented in Part I. The first six sections on performing routine operations with algebraic expressions and manipulating equations and formulas represent basic algebra that will be used throughout the calculus course. Even though this material is familiar, you should work through the problems to check your facility. Strive for accuracy and efficiency. The last three sections on graphing review the basics of analytical geometry. Understanding the relationship beween algebra and geometry is a prerequisite to understanding calculus. Your feeling for this correspondence will grow throughout the calculus course.

DIAGNOSTIC TEST: Over Part I

(Answers on page 187)

1. Simplify: $5(2\sqrt{7} - \sqrt{3}(2 - \sqrt{3})) + \sqrt{7}(4 + 3(\sqrt{7} - 1))$.

2. Add and simplify: $3[a - (b + 1)] - [2(b - a) + 1]$.

3. Perform the operations and simplify: $(a + \frac{1}{b}) \div (b + \frac{1}{a})$.

4. Solve for x: $1 - \frac{5}{x} = \frac{3}{4}$.

5. Sketch the graph of $y = 2 - x^3$.

6. Write the equation of the line through $(1,4)$ and $(3,-2)$.

7. Determine the value of k so that the slope of the line through $(2,3)$ and (k,k) equals $3/4$.

8. Find all values of x which satisfy each of the following statements:

    (i) $|2x - 3| \le 1$ (ii) $|x + 1| = 2|x - 4|$

9. Determine a value for k so that the point $(1,k)$ is the same distance from the line $x = -2$ as it is from the point $(2,0)$.

2

1. Combining Like Terms

Nearly every mathematical problem you will encounter in your science and mathematics courses will involve arithmetical computations in some way. The most basic of these require simple addition and subtraction of integers $(0, \pm 1, \pm 2, \ldots)$.

<u>Example 1.</u> Add $2a + 3b - 5 + 3a - 6b + 7$.

<u>Solution.</u> We group together the a terms, the b terms, and the constant terms,

$$(2a + 3a) + (3b - 6b) + (-5 + 7),$$

and combine the terms of each kind, by adding or subtracting their coefficients, to get

$$5a - 3b + 2.$$

<u>Example 2.</u> Add $3\sqrt{2} - 4\sqrt{3} + 2\sqrt{2} + 5\sqrt{3} - \sqrt{2} - 3\sqrt{3}$.

<u>Solution.</u> Combining $\sqrt{2}$ terms and $\sqrt{3}$ terms respectively, we get

$$4\sqrt{2} - 2\sqrt{3}.$$

Now it is easy, with a hand calculator, to find an approximate decimal value for this expression.

<u>Example 3.</u> Add $4x^2 - 2x + 1 - 3x^2 + 2x + 7 - 5x^2 - x + 3$.

<u>Solution.</u> In this problem we combine the x^2 terms, the x terms, and the constants to get

$$-4x^2 - x + 11.$$

3

EXERCISES

Add the expressions below in the manner illustrated in the preceding examples (combining like terms).

1. $4a + 2b + 7a - 5b$

2. $-2x + 7y - 5x + 2y$

3. $8\sqrt{3} + 4\sqrt{5} - 3\sqrt{3} + 2\sqrt{5} + 4\sqrt{3} - 5\sqrt{5}$

4. $8x^2 + 9x + 10 - 3x^2 - 6x - 5 + x^2 - x - 2$

5. $6a + 4b + c - 5a + b - 4c + 6a - c$

6. $2x + 3y + z + x - 4y - z + 7x + 5y - 4z$

7. $2 \log 2 + 7 + 3 \log 2 - 5 - 4 \log 2 - 1$

8. $4\sqrt{2} + 3\sqrt{3} - 5\sqrt{5} + \sqrt{3} - 2\sqrt{5} + \sqrt{2} + 7\sqrt{3}$

9. $4\pi + 2 - 3\pi - 5$

10. $3\pi + 4\sqrt{2} - 7\pi + 2\sqrt{2} - 3\sqrt{2} + 3\pi$

ANSWERS

1. $11a - 3b$ 2. $-7x + 9y$ 3. $9\sqrt{3} + \sqrt{5}$ 4. $6x^2 + 2x + 3$

5. $7a + 5b - 4c$ 6. $10x + 4y - 4z$ 7. $\log 2 + 1$

8. $5\sqrt{2} + 11\sqrt{3} - 7\sqrt{5}$ 9. $\pi - 3$ 10. $3\sqrt{2} - \pi$

---

2. Parentheses and Multiplication

In simplifying algebraic expressions that contain parentheses, first perform the operations shown within the parentheses and simplify the results; then carry out multiplications shown outside the parentheses; finally, combine like terms as in Section 1.

4

Example 1. Simplify $2(x^2 + 4x + 1) - 3(3x^2 - 2x + 1)$.

Solution. In this case, the expressions within the parentheses are already simplified, so we can carry out the multiplications immediately to get

$$2x^2 + 8x + 2 - 9x^2 + 6x - 3.$$

We then combine like terms to get

$$-7x^2 + 14x - 1.$$

A common pitfall is keeping track of signs when distributing a negative quantity throughout an expression. Thus, a common error in the problem above is to expand $-3(3x^2 - 2x + 1)$ into $-9x^2 - 6x + 3$ instead of $-9x^2 + 6x - 3$.

Example 2. Simplify $4(2 + 3\sqrt{3} - 5\sqrt{2}) + 2\sqrt{2}(-2 - \sqrt{2}) - \sqrt{3}(2 + \sqrt{3})$.

Solution. Carrying out the multiplications yields

$$8 + 12\sqrt{3} - 20\sqrt{2} - 4\sqrt{2} - 4 - 2\sqrt{3} - 3,$$

which simplifies to

$$1 + 10\sqrt{3} - 24\sqrt{2}.$$

Example 3. Simplify $-2(x + 3(y + 4)) + 2(-2 + 7(x - y))$.

Solution. We work "from the inside out"; that is, we simplify the quantities inside the parentheses first to obtain

$$-2(x + 3y + 12) + 2(-2 + 7x - 7y).$$

We then multiply to obtain

$$-2x - 6y - 24 - 4 + 14x - 14y.$$

Finally, collecting coefficients, we get

$$12x - 20y - 28.$$

EXERCISES

Simplify the expressions below in the manner illustrated in the preceding examples.

1. $(x + 2y + 7) - (7x - 2y + 1)$

2. $4(x^2 + 7x + 5) + 2(x^2 - x - 1) - (x^2 + 2x + 3)$

3. $4(\sin \theta + 2 \cos \theta + 1) - 3(\sin \theta + \cos \theta - 2)$

4. $2(x^2 + 3(x + 4)) - 5(2x + 3(x^2 + 1))$

5. $-(x^2 + 2x(x + 1) + 3) + 2(-5 + 2(x + 1) - 3x^2)$

6. $3[\sqrt{2} - 2(1 + 4\sqrt{2} - 5\sqrt{3}) + 4(\sqrt{2} + 3)] - 7(\sqrt{3} + \sqrt{2})$

7. $5a^x + 5(2 + 3a^x) - 7[4 + a^x(2 + (-3))] - 5(7 + a^x)$

8. $a(3 + 2b + c) + b(4a + 2) + c(2 - 3(a + 7))$

9. $x^2(2x + 4) + 7x[3 + 2(x - 4(x^2 + 1))]$

10. $x[1 + x(1 + x(1 + x))]$

ANSWERS

1. $-6x + 4y + 6$ 2. $5x^2 + 24x + 15$ 3. $\sin \theta + 5 \cos \theta + 10$

4. $-13x^2 - 4x + 9$ 5. $-9x^2 + 2x - 9$ 6. $-16\sqrt{2} + 23\sqrt{3} + 30$

7. $22a^x - 53$ 8. $3a + 2b - 19c + 6ab - 2ac$ 9. $-54x^3 + 18x^2 - 35x$

10. $x + x^2 + x^3 + x^4$

3. Addition and Subtraction of Fractions

The ability to manipulate fractions efficiently is an absolute necessity in carrying out routine work in calculus. In this section and the next we will consider typical computations with fractions that occur in problems throughout the course.

<u>Example 1.</u> Add $\frac{3}{4} + \frac{1}{4} - \frac{7}{4} - \frac{5}{4}$.

<u>Solution.</u> We may regard this as a sum of the form

$$3\left(\frac{1}{4}\right) + 1\left(\frac{1}{4}\right) - 7\left(\frac{1}{4}\right) - 5\left(\frac{1}{4}\right),$$

which, on combining like terms, yields $-8\left(\frac{1}{4}\right)$ or -2. Alternatively, since all the denominators are equal to 4, we may write the sum in the form

$$\frac{3 + 1 - 7 - 5}{4},$$

which equals $\frac{-8}{4}$ or -2.

<u>Example 2.</u> Add $\left[\frac{1}{3}(2)^3 - \frac{1}{2}(2)^2\right] - \left[\frac{1}{3}(-1)^3 - \frac{1}{2}(-1)^2\right]$.

<u>Solution.</u> First, we write this expression in the form

$$\left(\frac{8}{3} - \frac{4}{2}\right) - \left(-\frac{1}{3} - \frac{1}{2}\right).$$

Now, removing the parentheses,

$$\frac{8}{3} - \frac{4}{2} + \frac{1}{3} + \frac{1}{2}.$$

and adding like terms gives

$$\frac{9}{3} - \frac{3}{2} = 3 - \frac{3}{2} = \frac{6}{2} - \frac{3}{2} = \frac{3}{2}.$$

Example 3. Add $\frac{4}{7} + \frac{2}{3} - \frac{2}{9} + \frac{5}{6}$.

Solution. Here, the first step is to find a common denominator. The denominators in factored form are 7, 3, 3^2, and $2 \cdot 3$. From this, we see that the smallest common denominator is $2 \cdot 3^2 \cdot 7 = 126$. The above sum is therefore

$$\frac{4}{7} \cdot \frac{18}{18} + \frac{2}{3} \cdot \frac{42}{42} - \frac{2}{9} \cdot \frac{14}{14} + \frac{5}{6} \cdot \frac{21}{21} = \frac{4 \cdot 18 + 2 \cdot 42 - 2 \cdot 14 + 5 \cdot 21}{126}$$

$$= \frac{72 + 84 - 28 + 105}{126}$$

$$= \frac{233}{126} .$$

Example 4. Add $\frac{a}{b} + \frac{c}{d}$.

Solution. The common denominator is bd, so the steps are exactly the same as those in Example 3:

$$\frac{a}{b} \cdot \frac{d}{d} + \frac{b}{b} \cdot \frac{c}{d} \qquad \text{or} \qquad \frac{ad + bc}{bd} .$$

This simplification is often called "cross-multiplying" because of the pattern of multiplication:

$$\frac{a}{b} + \frac{c}{d} = \frac{ad + bc}{bd} .$$

Notice in the last example that $\frac{a}{b} + \frac{c}{d}$ is NOT equal to $\frac{a + c}{b + d}$ NOR is it equal to $\frac{a + c}{bd}$. Thus, for example, $\frac{2}{3} + \frac{1}{2} \neq \frac{2 + 1}{3 + 2}$ and

$\frac{2}{3} + \frac{1}{2} \neq \frac{2 + 1}{6}$; rather, $\frac{2}{3} + \frac{1}{2} = \frac{2 \cdot 2 + 3 \cdot 1}{3 \cdot 2} = \frac{4 + 3}{6} = \frac{7}{6} .$

EXERCISES

Perform the following operations:

1. $\dfrac{1}{3} - \dfrac{7}{3} + \dfrac{2}{3} + \dfrac{5}{3}$

2. $\left(\dfrac{27}{8} - 6\right) - \left(\dfrac{1}{8} - 2\right)$

3. $\dfrac{2}{3} + \dfrac{1}{2} - \dfrac{8}{3} + \dfrac{4}{3} - \dfrac{7}{2} + \dfrac{3}{2}$

4. $\dfrac{2}{3}\cos\theta - \dfrac{1}{2}\sin\theta + \dfrac{5}{6}\cos\theta + \dfrac{3}{2}\sin\theta$

5. $\dfrac{7}{2}a + \dfrac{2}{5}b - \dfrac{3}{2}c + \dfrac{1}{2}a - \dfrac{3}{4}b - \dfrac{1}{6}c$

6. $\left[\dfrac{3}{2}(1)^2 - \dfrac{1}{3}(1) + 2\right] - \left[\dfrac{3}{2}(-2)^2 - \dfrac{1}{3}(-2) + 2\right]$

7. $\left(\dfrac{2}{3} - \dfrac{1}{4}\right) + \left(\dfrac{1}{2} - \dfrac{1}{6}\right)$

8. $\dfrac{1}{2} + \left[\dfrac{3}{4} - \left(\dfrac{2}{3} + \dfrac{1}{2}\right) - \left(\dfrac{3}{2} - \dfrac{1}{3}\right)\right]$

9. $\dfrac{b}{c} + \dfrac{a}{d}$

10. $\dfrac{1}{x^2} + \dfrac{2}{xy} - \dfrac{3}{y}$

ANSWERS

1. $\dfrac{1}{3}$ 2. $-\dfrac{3}{4}$ 3. $-\dfrac{13}{6}$ 4. $\dfrac{3}{2}\cos\theta + \sin\theta$ 5. $4a - \dfrac{7}{20}b - \dfrac{5}{3}c$

6. $-\dfrac{11}{2}$ 7. $\dfrac{3}{4}$ 8. $-\dfrac{13}{12}$ 9. $\dfrac{bd + ac}{cd}$ 10. $\dfrac{y + 2x - 3x^2}{x^2 y}$

9

4. Multiplication and Division of Fractions

Computations with sums, products, quotients, and differences of fractions can get very messy; it is important to keep your work neatly displayed, to avoid confusion about the location of the numerator and denominator.

Example 1. Simplify $\dfrac{2 + 3 \cdot 4 + 2(-3)}{4(-3) + (-2)(-5) + 2 \cdot 4}$.

Solution. We simplify the numerator and denominator separately. Thus, the fraction becomes

$$\frac{2 + 12 - 6}{-12 + 10 + 8} = \frac{8}{6} = \frac{4}{3} .$$

Example 2. Simplify $\dfrac{-3(2 + 4x) + (7x + 2)x}{8x(-5 + 2x) + 3(5x - 3x^2) + 2}$.

Solution. Simplifying the numerator and denominator separately produces

$$\frac{-6 - 12x + 7x^2 + 2x}{-40x + 16x^2 + 15x - 9x^2 + 2} .$$

Combining like terms in the numerator and in the denominator gives

$$\frac{7x^2 - 10x - 6}{7x^2 - 25x + 2} .$$

If you recall that $\dfrac{a}{b} \cdot \dfrac{c}{d} = \dfrac{ac}{bd}$, you should have no difficulty with the next example.

Example 3. Simplify $\dfrac{2}{3} (x + \dfrac{5}{4} y) - \dfrac{1}{2} (\dfrac{2}{3} x - \dfrac{1}{2} y)$.

Solution. By multiplying, we get $\dfrac{2}{3} x + \dfrac{10}{12} y - \dfrac{1}{3} x + \dfrac{1}{4} y$. Combining like terms, we get

$$(\frac{2}{3} - \frac{1}{3})x + (\frac{10}{12} + \frac{1}{4})y$$

which simplifies to $\dfrac{1}{3} x + \dfrac{13}{12} y$.

In the next example remember that the quotient $\frac{a}{b} \div \frac{c}{d}$ is equal to the product $\frac{a}{b} \cdot \frac{d}{c} = \frac{ad}{bc}$.

Example 4. Simplify $\dfrac{\frac{2}{3} - \frac{1}{5}}{\frac{1}{4} - (-\frac{2}{3})}$.

Solution. Working with the numerator first, we get

$$\frac{2}{3} - \frac{1}{5} = \frac{10 - 3}{15} = \frac{7}{15} .$$

Simplifying the denominator, we get

$$\frac{1}{4} + \frac{2}{3} = \frac{3 + 8}{12} = \frac{11}{12} .$$

Thus the fraction equals

$$\frac{7}{15} \div \frac{11}{12} = \frac{7}{15} \cdot \frac{12}{11} = \frac{84}{165} = \frac{28}{55} .$$

EXERCISES

Perform the operations indicated in each of the following problems.

1. $\dfrac{19 - (-31)}{21 + (-12)}$

2. $\dfrac{(4 + 2 - 3) \cdot (7 + 1 - 3)}{(2 + 1 - 5) + (2 + (1 + 3))}$

3. $\dfrac{3 + 2 \cdot (4 + x)}{7 + (-2 - x)}$

4. $\dfrac{1}{3} [8 - 4 + \frac{1}{2}] - \frac{1}{2} [4 + 1 - \frac{4}{3}]$

5. $\dfrac{\frac{2}{3} + \frac{1}{2}}{4 + 1}$

6. $(\frac{2}{3} + \frac{1}{4}) \cdot (4 + 3)$

7. $\dfrac{\frac{1}{2} + \frac{2}{5}}{\frac{3}{5} - \frac{3}{2}}$

8. $\dfrac{\frac{3}{11} x + \frac{7}{11} x + \frac{2}{11}}{\frac{4}{9} x - \frac{7}{9} x - \frac{1}{9}}$

9. $\dfrac{\frac{2}{3} + \frac{4}{5} x}{\frac{8}{3} - \frac{2}{5} x}$

10. $\dfrac{2a + 4b + \frac{1}{2} (a + b)}{6a - 4b - \frac{1}{2} (a - b)}$

1. $\dfrac{50}{9}$ 2. $\dfrac{15}{4}$ 3. $\dfrac{11 + 2x}{5 - x}$ 4. $-\dfrac{1}{3}$ 5. $\dfrac{7}{30}$ 6. $\dfrac{77}{12}$

7. -1 8. $-\dfrac{18}{11} \left(\dfrac{5x + 1}{3x + 1} \right)$ 9. $\dfrac{5 + 6x}{20 - 3x}$ 10. $\dfrac{5a + 9b}{11a - 7b}$

---

5. First Degree Equations

The equations in this section make up the most common class of equations you will encounter in your science and calculus classes. In the following examples, notice that the various steps are simply applications of the techniques reviewed in the preceding sections.

Example 1. Solve for t: $80t + 2(100 - 60t)(-60) = 0$.

Solution. The first step is to expand the expression on the left side of the equation using the technique of Section 2,

$$80t + 2(100)(-60) + 2(-60t)(-60) = 0,$$

$$80t - 12{,}000 + 7200t = 0.$$

Then combine like terms to get

$$7280t - 12{,}000 = 0.$$

Now add 12,000 to each side of the equation

$$7280t = 12{,}000,$$

and divide each side by 7280 to get the answer:

$$t = \frac{12000}{7280} = \frac{1200}{728} = \frac{600}{364} = \frac{300}{182} = \frac{150}{91} \, .$$

Example 2. Solve for m:

$$(1.1 - m) + (2.3 - 2m)(2) + (3.8 - 3m)(3) + (5.0 - 4m)(4) = 0.$$

Solution. The general plan for solving such problems is first to get the unknown quantities on one side and the constants on the other side by adding or subtracting the same quantities from each side of the equation, and then to solve for the value of the variable by dividing by its coefficient. Thus, we get

$$[1.1 + (2.3)(2) + (3.8)(3) + (5.0)(4)] +$$

$$[(-1) + (-2)(2) + (-3)(3) + (-4)(4)]m = 0,$$

$$(-1 - 4 - 9 - 16)m = -[1.1 + 4.6 + 11.4 + 20],$$

$$-30m = -37.1,$$

$$m = \frac{37.1}{30} = \frac{371/10}{30} = \frac{371}{300}.$$

Example 3. The area A of a trapezoid is one-half the sum of the lengths of the bases b_1 and b_2 multiplied by the altitude h; symbolically,

$$A = \frac{1}{2}(b_1 + b_2)h.$$

Suppose that the area is 20, the altitude is 5, and one of the bases has length 5. Find the length of the other base.

Solution. Substituting A = 20, h = 5, and b_1 = 5 into the formula for the area of a trapezoid, we have

$$20 = \frac{1}{2}(5 + b_2) \cdot 5.$$

From this we eliminate fractions,

$$40 = 5(5 + b_2),$$

and divide both sides of the equation by 5 to get

$$8 = 5 + b_2, \text{ or equivalently } b_2 = 3.$$

These examples show how to solve first degree equations: (i) clear fractions and simplify by multiplying; (ii) add or subtract, if necessary, to manipulate the terms containing the unknown to one side of the equation and all other terms to the other side; and then (iii) solve for the value of the unknown by a simple division. The result may need further simplification (we'll have more to say about this in Sections 12, 13, and 14).

EXERCISES

Solve each of the following equations.

1. $2(3 + 2x) - 5(x + 1) = 3$

2. $8(2 + a) = 7 + 2(a - 3)$

3. $\dfrac{2x + 4}{5} = 7$

4. $\dfrac{y - 4}{5 + 2} = -1$

5. $\dfrac{2y + 3}{7} = \dfrac{3y + 1}{3}$

6. $\dfrac{1}{3} v + \dfrac{3}{4} = \dfrac{1}{2} (4 + v)$

7. $(\dfrac{1}{5} - \dfrac{2}{3} x) = 2(\dfrac{1}{3} + \dfrac{1}{5} x) + 4$

8. If an object weighs W_0 pounds on the surface of the earth, then it weighs

$$W = \frac{W_0}{(1 + \frac{1}{4000} r)^2}$$

pounds when it is r miles above the surface of the earth. If an object weighs 100 pounds when it is 400 miles above the surface of the earth, how much does this object weigh on the surface of the earth? (That is, solve for W_0 when $W = 100$ and $r = 400$.)

9. Suppose that the quantities R, r_1, and r_2 are related by the formula

$$\frac{1}{R} = \frac{1}{r_1} + \frac{1}{r_2}$$

Solve for r_2 when $R = 30$ and $r_1 = 20$.

10. Evaluate y' in the following formula when $x = 2$ and $y = 1$:

$$2xy + x^2y' - 3y^2y' = 0.$$

ANSWERS

1. $x = -2$ 2. $a = -5/2$ 3. $x = 31/2$ 4. $y = -3$ 5. $y = 2/15$

6. $v = -15/2$ 7. $x = -67/16$ 8. $W_0 = 121$ 9. $r_2 = -60$

10. $y' = -4$

6. Working with Formulas

A skill related to that of solving first degree equations, and a skill which is important in all science courses, is that of solving for one of the variables in a formula in terms of the other variables.

Example 1.   Solve for each of the variables P, r, and t in the formula

$$A = P(1 + rt).$$

Solution.   Treating P as the unknown and the other variables as known, we divide each side of the equation by $1 + rt$ to get

$$\frac{A}{1 + rt} = P, \quad \text{or equivalently,} \quad P = \frac{A}{1 + rt}.$$

To solve for r (and t) we first carry out the multiplication indicated in the original formula, then isolate the unknowns. This yields

$$A = P + Prt,$$

$$A - P = Prt, \quad \text{or equivalently,} \quad Prt = A - P,$$

15

from which it follows that

$$r = \frac{A - P}{Pt} \quad \text{and} \quad t = \frac{A - P}{Pr} \,.$$

Example 2. Solve for r in the formula $V = \pi r^2 h$.

Solution. Dividing each side of the equation by πh we get

$$\frac{V}{\pi h} = r^2, \quad \text{or equivalently,} \quad r^2 = \frac{V}{\pi h} \,.$$

Taking the square root of each side, we get

$$r = \sqrt{\frac{V}{\pi h}} \,.$$

We take only the positive square root, because the original formula represented the volume of a right circular cylinder whose radius r must be positive.

As we have seen, when a formula contains fractions, the first step is to eliminate these fractions by multiplying each side of the equation by a common multiple of all the denominators.

Example 3. Solve for r_1 in the formula $\frac{1}{R} = \frac{1}{r_1} + \frac{1}{r_2}$.

Solution. To begin, we multiply both sides of the equation by Rr_1r_2 to get

$$r_1r_2 = Rr_2 + Rr_1.$$

Next, by collecting terms that contain r_1, we get

$$r_1r_2 - Rr_1 = Rr_2$$

or

$$(r_2 - R)r_1 = Rr_2,$$

and, finally,

$$r_1 = \frac{Rr_2}{r_2 - R} \,.$$

16

EXERCISES

Solve each formula (equation) for each letter that follows it.

1. $\dfrac{V_1}{V_2} = \dfrac{P_2}{P_1}$; V_1, P_2, P_1

2. $s = \dfrac{1}{2} gt^2$; g

3. $C = \dfrac{5}{9}(F - 32)$; F

4. $A = \dfrac{1}{2} h(b_1 + b_2)$; h, b_1

5. $F = \dfrac{Wv^2}{gr}$; W, g

6. $s = \dfrac{a}{1 - r}$; a, r

7. $I = \dfrac{E}{R + nr}$; E, R, r

8. $\dfrac{1}{u} + \dfrac{1}{v} = \dfrac{1}{f}$; f, u

9. $F = k\dfrac{m_1 m_2}{d^2}$; m_1, d^2

10. $A = \pi x^2 + xy$; y

ANSWERS

1. $V_1 = P_2 V_2 / P_1$, $P_2 = V_1 P_1 / V_2$, $P_1 = P_2 V_2 / V_1$ 2. $g = 2s/t^2$

3. $F = \dfrac{9}{5} C + 32$ 4. $h = 2A/(b_1 + b_2)$, $b_1 = (2A/h) - b_2 = (2A - b_2 h)/h$

5. $W = Fgr/v^2$, $g = Wv^2/Fr$ 6. $a = s(1 - r)$, $r = 1 - a/s = (s - a)/s$

7. $E = I(R + nr)$, $R = E/I - nr = (E - Inr)/I$, $r = (E/I - R)/n = (E - RI)/In$

8. $f = uv/(u + v)$, $u = fv/(v - f)$ 9. $m_1 = Fd^2/km_2$, $d^2 = km_1 m_2/F$

10. $y = (A - \pi x^2)/x$

7. Graphing

A rectangular coordinate system in the plane identifies points by ordered pairs of numbers. In the system shown here the points corresponding to (1,2), (-2,-3), and (4,-2) are as plotted.

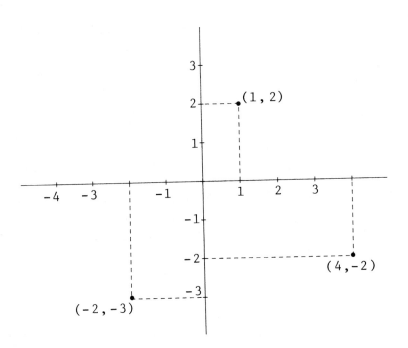

The most important application of this correspondence is to graphing.

Example 1. Sketch the graph of $y = x^2$.

Solution. The problem is to plot those points in the plane whose coordinates (x,y) satisfy $y = x^2$. The usual procedure is to make a table, substituting various values of x into the equation $y = x^2$ in order to compute the corresponding y-value which satisfies the given equation. Plotting the points in the table and sketching a smooth curve through them produces the graph shown.

x	y $(= x^2)$
±2	4
±1	1
±1/2	1/4
0	0

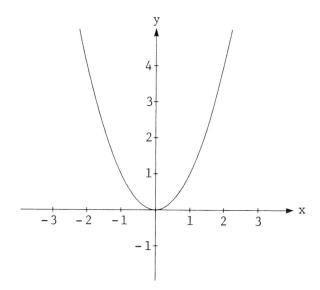

Example 2. Sketch the graph of y = x(4 − x).

Solution.

x	y
−1	−5
0	0
1	3
2	4
3	3
4	0
5	−5

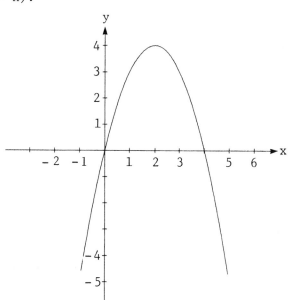

Example 3. Sketch the graph of $y = x + \dfrac{1}{x}$.

Solution. From the table

x	1	2	3	4	...
y	2	$2\frac{1}{2}$	$3\frac{1}{3}$	$4\frac{1}{4}$	...

we see that for large values of x, x and y are approximately equal; and from

19

the table

x	$\frac{1}{2}$	$\frac{1}{3}$	$\frac{1}{4}$	$\cdots$
y	$2\frac{1}{2}$	$3\frac{1}{3}$	$4\frac{1}{4}$	$\cdots$

we see that as x approaches zero from the positive side, y becomes arbitrarily large. Also, note that $(-x) + \frac{1}{(-x)} = -(x + \frac{1}{x}) = -y$, so in the tables above, if x is replaced by $-x$, then y is replaced by $-y$. (For example, $(-3, -3\frac{1}{3})$ is also a point on the graph.) Note that there is a break in the "continuity" of the graph at $x = 0$, i.e., the two "halves" are not connected.

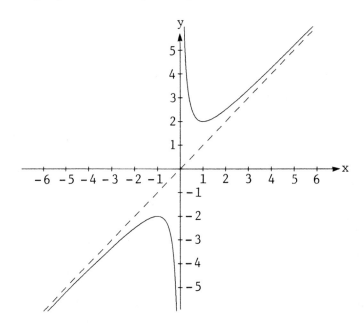

One of the most important requirements in producing good graphs is to plot points accurately according to a consistent scale (the scale can differ on the separate axes) and to keep the work neatly organized.

EXERCISES

Sketch the graphs for the following equations.

1. $y = 2x$

2. $y = -x + 2$

3. $y = 8x - x^2$

4. $y = x^3$

5. $y = x^2 + 3x + 4$

6. $y = 3 - x - x^2$

7. $y = x^3 - 9x$

8. $y = 1/x$

9. $2x + 3y = 6$

10. $y^2 = x$

ANSWERS

1.
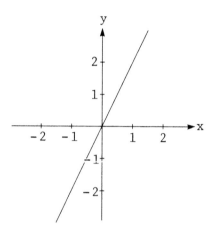

2.
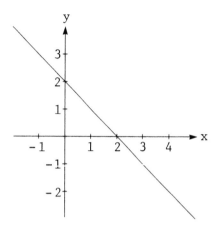

3.
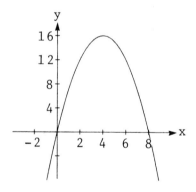

4.
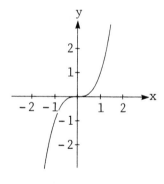

5.
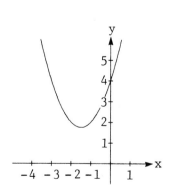

6.
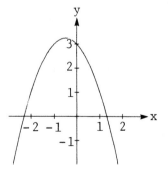

21

7.

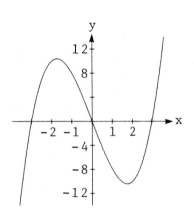

8.

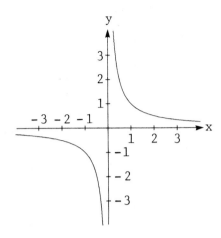

9.

10.

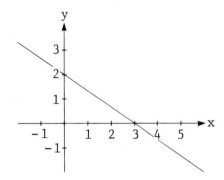

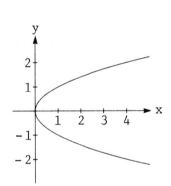

---

8. Lines

A point in the plane is on the graph of y = mx if and only if its y-coordinate is m times its x-coordinate. This implies that the graph of y = mx is a straight line through the origin in which a change of (positive) one unit in x produces a change of m units in y. The constant m is called the slope of the line. Some lines corresponding to different values of m are shown below.

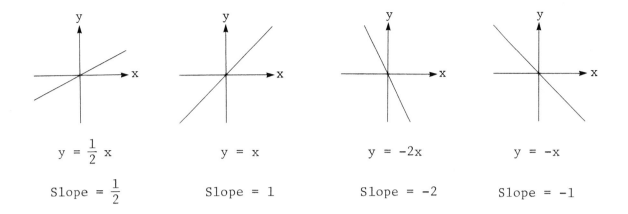

$$y = \frac{1}{2} x$$

$$y = x$$

$$y = -2x$$

$$y = -x$$

Slope $= \frac{1}{2}$

Slope $= 1$

Slope $= -2$

Slope $= -1$

Positive values of m correspond to positive slopes: their graphs rise from the lower left to the upper right. Negative values of m correspond to negative slopes: their graphs fall from the upper left to the lower right.

The graph of $y = mx + b$ is simply the graph of $y = mx$ moved up or down b units. For example, the graph of $y = 2x + 3$ is the graph of $y = 2x$ moved three units upward, and the graph of $y = 2x - 2$ is the graph of $y = 2x$ moved two units downward.

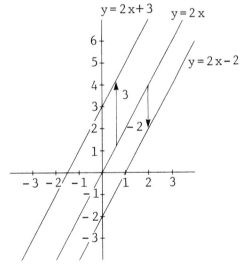

The slope of each of these is equal to 2. In general $y = mx + b$ is a straight line with slope m and y-intercept equal to b. This is called the *slope-intercept form* for the equation of a straight line; it can be used to

23

sketch graphs of straight lines quickly (making the cumbersome plotting methods of the last section unnecessary for straight lines).

Example 1. Graph the equation $2x - 3y - 6 = 0$.

Solution. Begin by putting the equation in the slope-intercept form (that is, solve for y):

$$y = \frac{2}{3} x - 2.$$

In this form we know the equation is that of a straight line with y-intercept equal to -2 and slope equal to $2/3$ (a change of 1 in x corresponds to a change of $2/3$ in y, or equivalently, a change of 3 in x yields a change of 2 in y). The graph is shown below.

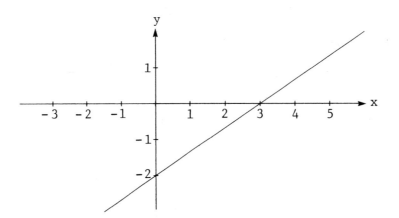

Observe that any equation of the form $Ax + By + C = 0$ can be put into the slope-intercept form provided $B \neq 0$ (simply solve for y). Equations of this form are called *linear equations* because they correspond geometrically to straight lines. Note that the equation $Ax + By + C = 0$ is a straight line also when $B = 0$ (and $A \neq 0$); namely, it represents the vertical line through $(-\frac{C}{A}, 0)$.

Although the slope-intercept form is very helpful in graphing linear equations, there is another way of thinking about lines and their equations which is more efficient for most applications. Before presenting this method

24

recall that if $P = (x_1, y_1)$ and $Q = (x_2, y_2)$ are any two points in the plane with $x_1 \neq x_2$, then the slope of the line which passes through P and Q is

$$\frac{y_2 - y_1}{x_2 - x_1} \quad \left(= \frac{y_1 - y_2}{x_1 - x_2} \right) .$$

This follows from the earlier definition (a change of one in x produces a change of m in y) since (see the figure) corresponding sides of similar triangles are proportional: that is,

$$\frac{y_2 - y_1}{x_2 - x_1} = m \quad (= \frac{m}{1}) .$$

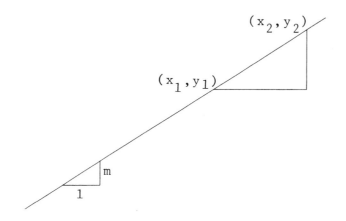

Now suppose that $P = (x_1, y_1)$ is a point on a line L whose slope is m. Then an arbitrary point $R = (x, y)$ in the plane, but different from P, is on L if and only if the slope of the line through R and P is equal to m; that is, if and only if

$$\frac{y - y_1}{x - x_1} = m,$$

$$y - y_1 = m(x - x_1) .$$

This is called the *point-slope form* for the equation of a line. To apply this method to find the equation of a line, you need to know a *point* on the line and the *slope* of the line. This procedure for finding the equation of a line is preferred in calculus, so the reasoning should be studied carefully. The next example offers a more concrete instance of the pertinent ideas.

Example 2. Find the equation of the line which passes through the points
A = (2,-1) and B = (-3,2).

Solution. We are given a point on the line (we may choose either A or B); we can apply the point-slope reasoning provided we can determine the slope of the line. We are given that the line passes through A and B, and therefore its slope is $\frac{2 - (-1)}{-3 - 2}$ = $-\frac{3}{5}$. If P = (x,y) is any point different from A, then P is on the line through A and B if and only if the slope of AP equals the slope of AB; that is, if and only if

$$\frac{y - (-1)}{x - 2} = -\frac{3}{5} .$$

It follows that the equation of L is

$$5(y + 1) = -3(x - 2),$$

which simplifies to

$$3x + 5y = 1.$$

This equation in slope-intercept form is

$$y = -\frac{3}{5} x + \frac{1}{5} .$$

EXERCISES

Put the equations below into slope-intercept form and sketch their graphs.

1. x - 2y + 2 = 0   2. 3x - 2y + 6 = 0   3. 2x + 3y - 1 = 0

Use the point-slope method to find the equation of each of the following lines.

4. Line through (4,1) with slope 2

5. Line through (-1,2) and (3,1)

6. Line through (4,7) parallel to y = 3x + 7

7. Suppose that L is a line of slope m, m ≠ 0. It can be shown that the slope of any line perpendicular to L has slope -1/m (negative reciprocal). Use this fact to find the equation of the line through (4,7) which is perpendicular to $y = \frac{2}{3} x + 1$.

8. Write the equation of the line through (4,2) and (4,3).

9. Where does the line through (2,5) and (5,-4) intersect the x-axis? (Hint: Write the equation of the line, set y = 0, and solve for x.)

10. Where does the line through (2,5) and (5,9) intersect the vertical line x = 7?

ANSWERS

1. $y = \frac{1}{2} x + 1$

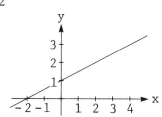

2. $y = \frac{3}{2} x + 3$

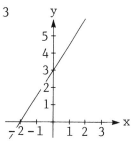

3. $y = -\frac{2}{3} x + \frac{1}{3}$

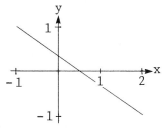

4. $y - 1 = 2(x - 4)$

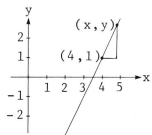

5. $y - 2 = -\frac{1}{4} (x + 1)$, or equivalently, $y - 1 = -\frac{1}{4} (x - 3)$

6. $y - 7 = 3(x - 4)$ 7. $y - 7 = -\frac{3}{2} (x - 4)$ 8. $x = 4$ 9. $x = \frac{11}{3}$

10. $(7, \frac{35}{3})$

9. Linear Inequalities, Absolute Value, and Distance

If a and b are real numbers and a < b, then for any real number c,

$$a + c < b + c, \quad \text{and} \quad a - c < b - c.$$

Also, if c is a positive number,

$$ac < bc, \quad \text{and} \quad a/c < b/c,$$

whereas if c is negative the inequalities reverse, and we have

$$ac > bc, \quad \text{and} \quad a/c > b/c.$$

Example 1. Find all real numbers x which satisfy the linear inequality

$$3x + 2 < 5x - 7.$$

Solution. We proceed exactly as if we were solving linear equations, taking variables to the left side and constants to the right side. Thus, subtracting 5x + 2 from each side of the inequality yields -2x < -9, and dividing by -2 gives x > 9/2. Geometrically, these numbers correspond to the points to the right of 9/2:

The *absolute value* of a real number a, denoted by $|a|$, is the magnitude of the distance from 0 to a. Thus, for example,

$$|2| = 2,$$
$$|-2| = 2,$$
$$|-7| = 7,$$

and, in general,

$$|a| = \begin{cases} a & \text{if } a \text{ is positive} \\ 0 & \text{if } a \text{ is zero} \\ -a & \text{if } a \text{ is negative.} \end{cases}$$

Example 2. Find all real numbers x for which $|2x + 7| = 3$.

Solution. There are two numbers that are 3 units from 0, namely 3 and -3. Thus, either $2x + 7 = 3$, or $2x + 7 = -3$. In the first case, $x = \frac{3 - 7}{2} = -2$, and in the second case $x = \frac{-3 - 7}{2} = -5$. It follows that there are two solutions to the given equation: $x = -2$ and $x = -5$.

The magnitude of the distance between two numbers a and b is $|a - b|$ (or equivalently, $|b - a|$). Conversely, $|a - b|$ can be interpreted as the (magnitude of the) distance between a and b. This geometrical interpretation can often simplify solving equations with absolute values.

Example 3. Find all real numbers x which satisfy each of the following statements: a) $|x - 3| = 2$

b) $|x - 3| < 2$.

Solution. a) Proceeding as in Example 2, either $x - 3 = 2$ or $x - 3 = -2$. From these equations we get two solutions: $x = 5$, $x = 1$.

Alternatively, we wish to find all numbers x whose distances from 3 are equal to 2. These numbers are easily seen to be $x = 1$ and $x = 5$.

b) It is easiest to interpret the problem as asking us to find all numbers x whose distances from 3 are less than 2. The line graph in part (a) shows that these numbers are greater than 1 and less than 5.

29

Alternatively, we can reason as follows: (x - 3) represents a quantity whose distance from zero is less than 2. It follows that

$$-2 < x - 3 < 2.$$

Adding 3 to each of these inequalities gives the same solution as before, namely,

$$1 < x < 5.$$

Return now to the consideration of Example 2: find all x which satisfy

$$\left|2x + 7\right| = 3.$$

Dividing each side by 2 (the coefficient of x) yields

$$\left|x + \frac{7}{2}\right| = \frac{3}{2}$$

or, equivalently,

$$\left|x - \left(-\frac{7}{2}\right)\right| = \frac{3}{2}.$$

When the problem is put into this form, we can interpret it as asking us to find those numbers x whose distances from -7/2 are 3/2.

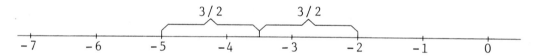

The line graph shows these points are x = -5 and x = -2.

Example 3. Find all real numbers x for which $\left|3x - 10\right| > 2$.

Solution. One approach is to factor 3 from the left side and put the inequality in the alternate form

$$\left|x - \frac{10}{3}\right| > \frac{2}{3}$$

In this form, we wish to find all numbers x whose distances from 10/3 are greater than 2/3.

From the line graph we see that these numbers are either less than 8/3 or greater than 4.

Another approach is to argue that if $|3x - 10| > 2$ then either $3x - 10 > 2$ or $3x - 10 < -2$. (It would *not* be correct to express this as $2 < 3x - 10 < -2$.) The first inequality is equivalent to $x > 4$ and the second to $x < 8/3$.

A third approach is to find those x's for which $|3x - 10| \leq 2$. This is equivalent to $-2 \leq 3x - 10 \leq 2$, which in turn is equivalent to $8/3 \leq x \leq 4$. It follows that the numbers for which $|3x - 10| > 2$ are those, as before, which are not between 8/3 and 4 inclusive.

Absolute values satisfy the following rules:

(1) $|a| = |-a|$,

(2) $|ab| = |a||b|$,

(3) $\left|\dfrac{a}{b}\right| = \dfrac{|a|}{|b|}$, $b \neq 0$,

(4) $|a + b| \leq |a| + |b|$ (the triangle inequality).

We conclude this section by considering the problem of determining the distance between two points in the plane. Let $P = (x_1, y_1)$ and $Q = (x_2, y_2)$ be given points. The distance between their y-coordinates is $|y_2 - y_1|$; the distance between their x-coordinates is $|x_2 - x_1|$.

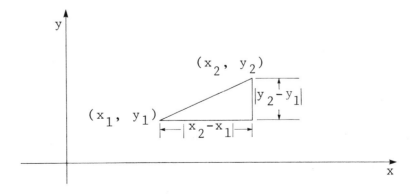

Therefore, by the Pythagorean Theorem, the distance d between P and Q is

31

$$d = \sqrt{|y_2 - y_1|^2 + |x_2 - x_1|^2}$$

$$= \sqrt{(y_2 - y_1)^2 + (x_2 - x_1)^2}\ .$$

EXERCISES

1. Show that if $0 < x < y$, then $1/y < 1/x$.

2. Show that if $0 < x < 1$, then $x^2 < x$, whereas if $x > 1$, then $x^2 > x$.

3. If $x < y$, does it follow that $x^2 < y^2$? If $0 < x < y$, does it follow that $x^2 < y^2$?

Find all numbers x which satisfy the following statements. Do each of the absolute values problems in two ways.

4. $4x + 3 > 0$

5. $3x + 2 < 4x - 3$

6. a) $|x - 2| = 5$,   b) $|x - 2| < 5$

7. a) $|4x + 12| = 8$,   b) $|x + 3| < 2$

8. $|x - 2| > 1$

9. $|2x - 3| < 5$

10. If $|x| < 2$ and $|y| < 3$, use the triangle inequality and the rules of absolute values to prove that $|5x^2 - 7y| < 41$.

11. Compute the distance between each of the following pairs of points.

   a) $(2,3)$, $(5,7)$ b) $(-4,2)$, $(5,-1)$ c) $(7,2)$, $(5,2)$

ANSWERS

1. Divide by xy.   2. Multiply by x.   3. a) NO. Take $x = -2$, $y = 1$.
b) YES. $0 < x < y$ implies $x^2 < xy$ and $xy < y^2$ from which it follows that $x^2 < y^2$.   4. $x > -3/4$ 5. $x > 5$ 6. a) $x = -3$, $x = 7$
b) $-3 < x < 7$ 7. a) $x = -5$, $x = -1$;   b) $-5 < x < -1$ 8. $x < 1$, $x > 3$
9. $-1 < x < 4$ 10. $|5x^2 - 7y| \le |5x^2| + |-7y| =$
$5|x|^2 + 7|y| < 5 \cdot 2^2 + 7 \cdot 3 = 41$ 11. a) 5; b) $3\sqrt{10}$;   c) 2

PART II: Algebra for Derivatives

The first major concept in calculus is the derivative. The
early introduction of this important concept imposes a strict or-
ganizational structure: (a) functions, (b) limits, (c) computing
derivatives using the definition, (d) computing derivatives using
the rules of differentiation. Algebra is used extensively in (b),
(c), and (d) to simplify complicated algebraic expressions. The
skills that you should review for this work include the multiplication of polynomials, factoring, and simplifying expressions containing radicals. For the definition of the derivative, you will
need to understand the use of functional notation and know how
to interpret functions geometrically. In addition, you will need
to know how to work with functions containing positive and negative fractional exponents because the derivatives of such functions
are the first to be considered.

(Answers on page 187)

1. Express $\left(-\dfrac{x^2}{y}\right)^3 \left(\dfrac{x^{-1}}{2y^2}\right)^2$ as a quotient with no negative exponents.

2. Add and simplify: $2x(x + 1)^{-2} - 2x^2(x + 1)^{-3}$.

3. One light-year is the distance light travels in one year. Use scientific notation to make an estimate for the number of miles in a light year. (Assume that the speed of light is 186,000 miles per second.)

4. Expand by multiplication and simplify: $(x + 2y)^2 - (2x - y)^2$.

5. Simplify: $\dfrac{x^2 - 9}{6x^2 - x - 2} \div \dfrac{x^2 + x - 6}{3x^2 - 8x + 4}$.

6. Add and simplify: $2x\sqrt{x^2 + 4} - \dfrac{x^3}{\sqrt{x^2 + 4}}$.

7. Simplify: $\dfrac{\sqrt[3]{16x^2y^5}\ \sqrt[3]{9x^4y^2}}{\sqrt[3]{6xy^2}}$.

8. If $f(x) = \dfrac{1}{1 - x}$, show that $f(1 + x) = -f(1 - x)$.

9. Solve for x: $y = \dfrac{2x + 1}{x + 1}$.

10. Integer Exponents

Recall the following notational conventions: if a is different from zero and n is a positive integer, then

$$a^n = \underbrace{a \cdot a \cdots a}_{n \text{ times}}$$

$$a^{-n} = \frac{1}{a^n}$$

$$a^0 = 1.$$

Three rules of exponents allow you to simplify complicated expressions containing exponents,

(1) $a^n a^m = a^{n+m}$,   n and m integers, $a \neq 0$.

(2) $(a^n)^m = a^{nm}$,   n and m integers, $a \neq 0$

(3) $(ab)^n = a^n b^n$,   n an integer, $a \neq 0$, $b \neq 0$.

These rules are easily remembered if you check them out with some examples.

$$a^3 a^4 = (a \cdot a \cdot a)(a \cdot a \cdot a \cdot a) = (a \cdot a \cdot a \cdot a \cdot a \cdot a \cdot a) = a^7 = a^{3+4}$$

$$a^3 a^{-4} = \frac{a^3}{a^4} = \frac{a \cdot a \cdot a}{a \cdot a \cdot a \cdot a} = \frac{1}{a} = a^{-1} = a^{3-4}$$

$$a^3 a^{-3} = \frac{a^3}{a^3} = \frac{a \cdot a \cdot a}{a \cdot a \cdot a} = 1 = a^0 = a^{3-3}$$

$$(a^3)^4 = (a \cdot a \cdot a)(a \cdot a \cdot a)(a \cdot a \cdot a)(a \cdot a \cdot a) = a^{12} = a^{3 \cdot 4}$$

$$(a^{-3})^4 = \left(\frac{1}{a^3}\right)^4 = \frac{1}{a^{12}} = a^{-12} = a^{(-3) \cdot 4}$$

$$(ab)^3 = (ab)(ab)(ab) = (a \cdot a \cdot a)(b \cdot b \cdot b) = a^3 b^3$$

The following rules are special cases of rule (1). (Compare the first and last expressions in these equations.)

$$(4) \quad \frac{a^n}{a^m} = a^n \left(\frac{1}{a^m} \right) = a^n a^{-m} = a^{n-m}$$

$$(5) \quad \frac{a^n}{a^m} = a^{n-m} = a^{-(m-n)} = \frac{1}{a^{m-n}} \, .$$

Another useful rule is a special case of rule (3):

$$(6) \quad \left(\frac{a}{b} \right)^n = \left[a \left(\frac{1}{b} \right) \right]^n = a^n \left(\frac{1}{b} \right)^n = a^n \frac{1}{b^n} = \frac{a^n}{b^n} \, .$$

Example 1. Express $\dfrac{ab^{-2}}{c^{-3}}$ as a quotient with no negative exponents.

Solution. $\dfrac{ab^{-2}}{c^{-3}} = \dfrac{a \left(\frac{1}{b^2} \right)}{\left(\frac{1}{c^3} \right)} = \dfrac{a}{b^2} \cdot \dfrac{c^3}{1} = \dfrac{ac^3}{b^2} \, .$

Example 1 illustrates the general principle that any term x^n in a product or quotient can be moved from the numerator to the denominator or vice versa by changing the sign of the exponent. This principle, together with repeated application of the rules of exponents (1) – (6), allows you to simplify complicated products by eliminating negative exponents and combining factors having the same base.

Example 2. Simplify $\left(\dfrac{a^4 b^{-5}}{a^{-2} b^2} \right)^3 \left(\dfrac{a^{-3} b^{-8}}{a^{-5} b^{-6}} \right)^{-4} .$

Solution. We aim to eliminate negative exponents and to use as few exponents as possible. One approach is first to eliminate negative exponents within the parentheses:

$$\left(\frac{a^4 b^{-5}}{a^{-2} b^2}\right)^3 \left(\frac{a^{-3} b^{-8}}{a^{-5} b^{-6}}\right)^{-4} = \left(\frac{a^4 a^2}{b^5 b^2}\right)^3 \left(\frac{a^5 b^6}{a^3 b^8}\right)^{-4}$$

$$= \left(\frac{a^6}{b^7}\right)^3 \left(\frac{a^2}{b^2}\right)^{-4}$$

$$= \frac{a^{18}}{b^{21}} \cdot \frac{a^{-8}}{b^{-8}}$$

$$= \frac{a^{10}}{b^{13}}$$

Alternatively, we could proceed:

$$\left(\frac{a^4 b^{-5}}{a^{-2} b^2}\right)^3 \left(\frac{a^{-3} b^{-8}}{a^{-5} b^{-6}}\right)^{-4} = (a^6 b^{-7})^3 (a^2 b^{-2})^{-4}$$

$$= a^{18} b^{-21} a^{-8} b^8$$

$$= a^{10} b^{-13}$$

$$= \frac{a^{10}}{b^{13}} .$$

<u>Example 3.</u> Simplify $\dfrac{(x+1)^{-1}}{x^{-1}+1}$.

<u>Solution.</u> We first rewrite to eliminate negative exponents, to obtain

$$\frac{\dfrac{1}{x+1}}{\dfrac{1}{x}+1} .$$

Next, we simplify the denominator (the numerator is already a simple fraction)

to get

$$\frac{\dfrac{1}{x+1}}{\dfrac{1+x}{x}} .$$

37

The resulting quotient is equivalent to

$$\frac{1}{x+1} \cdot \frac{x}{1+x} = \frac{x}{(1+x)^2} \ .$$

Many algebraic mistakes are due to incorrect handling of exponents. In working with exponents, note that:

$$(3x)^2 \text{ is NOT equal to } 3x^2,$$

$$ab^2 \text{ is NOT equal to } (ab)^2,$$

$$(a+b)^2 \text{ is NOT equal to } a^2 + b^2,$$

$$a^{-1} + b^{-1} \text{ is NOT equal to } (a+b)^{-1}.$$

When you are in doubt about the validity of an exponential manipulation (or any other algebraic property), you should check it out by substituting particular values for the letters. For example, to see that

$$a^{-1} + b^{-1} \neq (a+b)^{-1}$$

consider the special case a = 2, b = 3. Note that $2^{-1} + 3^{-1} = \frac{1}{2} + \frac{1}{3} = \frac{5}{6}$, whereas $(2+3)^{-1} = 5^{-1} = \frac{1}{5}$. Of course, a single counterexample is enough to disprove a proposed identity; however, it should be understood that even lots of favorable special cases will not prove the identity for all possible values of the letters.

SCIENTIFIC NOTATION

Exponential notation provides an efficient way of writing very large or very small numbers. You have no doubt seen this notation in your experience with hand calculators, and you will certainly use it often when you study science. A number is in *scientific form* when it is written in the form

$$k \times 10^n$$

where $1 \le k < 10$ and n is an integer.

Example 4. Write each of the following numbers in scientific form:

    a) 365

    b) 93,000,000

    c) .00032

Solution. a) $365 = 3.65 \times 10^2$

        b) $93,000,000 = 9.3 \times 10^7$

        c) $.00032 = 3.2 \times 10^{-4}$

One of the most useful applications of scientific notation is in providing estimates for the order of magnitude of a complicated computation.

Example 5. Give an estimate for the value of

$$\frac{(17,200,000,000)(.00000957)}{(.003)(82,000)} .$$

Solution. We begin by expressing each of the terms in scientific form

$$\frac{(1.72 \times 10^{10})(9.57 \times 10^{-6})}{(3 \times 10^{-3})(8.2 \times 10^4)} .$$

By the rules of exponents, this is equal to

$$\frac{(1.72)(9.57)}{(3)(8.2)} \times 10^3 .$$

To estimate the quotient on the left, we have

$$\frac{(1.72)(9.57)}{(3)(8.2)} \cong \left[\frac{1.72}{8.2}\right] \left(\frac{10}{3}\right) \cong \left(\frac{2}{10}\right)\left(\frac{10}{3}\right) = \frac{2}{3} .$$

Thus, an estimate for the original problem is

$$\frac{2}{3} \times 10^3 \cong 667 .$$

(With a hand calculator we get

$$\frac{(1.72)(9.57)}{(3)(8.2)} \cong .669$$

which gives the more precise answer of 669 to the original problem.)

Example 6. Given that $2^{10} = 1024 \cong 10^3$, give an estimate in scientific form to $2^{21,701} - 1$ (this number is known to be a prime number).

Solution.

$$2^{21,701} = 2(2^{21,700}) = 2(2^{10})^{2,170}$$

$$\cong 2(10^3)^{2,170} = 2 \times 10^{6,510}.$$

Thus, $2^{21,701} - 1$ has approximately 6,510 digits when expressed in decimal notation (the actual number of digits is 6,533).

EXERCISES

Simplify the following expressions.

1. $(2^2 \cdot 3^3 \cdot 5 \cdot 7)(2 \cdot 3^4 \cdot 5 \cdot 11)$

2. $(3a^2 b)^2 (2ab^4)^3$

3. $\dfrac{8a^2 b^3 c^0}{12ab^5 c}$

4. $\dfrac{x^{-1} y^{-4} z^3}{x^3 y^{-2} z^5}$

5. $\left(\dfrac{21A^2 B^{-4}}{15A^3 B^5}\right)^2 \left(\dfrac{-14AB^{-2}}{3A^5 B^2}\right)^{-3}$

6. $\dfrac{(2^n \cdot 3^m)^2 \, 2^{3n}}{3^m}$, n and m integers

Exercises 7-10, containing both sums and products, are more typical of the kinds of problems which are common in calculus. In each case, eliminate the negative exponents, perform the operations, and simplify.

7. $2x^{-1} + (2x)^{-1}$

8. $(x^{-1} + y^{-1})^{-1}$

9. $(x + 1)^{-1} - x(x + 1)^{-2}$

10. $2x(x^3 + 2)^{-2} - 2x^2 \cdot 3x^2 (x^3 + 2)^{-3}$

Write each of the following numbers in scientific form.

11. 840

12. 210,000,000,000,000

13. .047

14. .0000000083

Use scientific notation to give estimates for each of the following quantities, then use your hand calculator to give more precise values and express the result in scientific form.

15. $1/3100$

16. $\dfrac{3.5 \times 10^7}{2.8 \times 10^3}$

17. $\dfrac{(78,000)(.0000065)}{(2100)(.0002)}$

18. $\dfrac{(3.9 \times 10^{-19}) \times (8.6 \times 10^4)}{(9.3 \times 10^{21}) \times (2.7 \times 10^{-29})}$

19. Given that $2^{10} \simeq 10^3$, give an estimate in scientific form to the quantity 2^{x^2} when $x = 25$. (The expression k^{m^n} is understood to mean $k^{(m^n)}$ rather than $(k^m)^n = k^{mn}$.)

ANSWERS

1. $2^3 \cdot 3^7 \cdot 5^2 \cdot 7 \cdot 11$ 2. $72a^7 b^{14}$ 3. $2a/3b^2 c$ 4. $1/x^4 y^2 z^2$

5. $\dfrac{-3^3 \cdot A^{10}}{2^3 \cdot 5^2 \cdot 7 \cdot B^6}$ 6. $2^{5n} 3^m$ 7. $5/2x$ 8. $xy/(x + y)$ 9. $\dfrac{1}{(x + 1)^2}$

10. $\dfrac{4x - 4x^4}{(x^3 + 2)^3}$ 11. 8.4×10^2 12. 2.1×10^{14} 13. 4.7×10^{-2}

14. 8.3×10^{-9} 15. 3.2258×10^{-4} 16. 1.25×10^4 17. 1.207

18. 1.3357×10^{-7} 19. 32×10^{186}, or equivalently, 3.2×10^{187}

11. Multiplication of Polynomials

The multiplication of (ax + b) by (cx + d) can be carried out in two steps:

$$(ax + b)(cx + d) = ax(cx + d) + b(cx + d)$$

$$= acx^2 + adx + bcx + bd$$

$$= acx^2 + (ad + bc)x + bd.$$

However, such expansions are customarily written in a single step by following this procedure (the arrows represent multiplications):

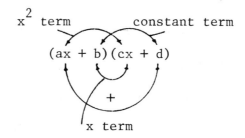

Example 1. Expand by multiplication and simplify:

$$(2x + 5)(3x - 4) + (-3x + 1)(3x + 4) + (-3x + 1)(2x + 5).$$

Solution. Using the short cut shown in the diagram, we have

$$(6x^2 + 7x - 20) + (-9x^2 - 9x + 4) + (-6x^2 - 13x + 5),$$

and, collecting like terms, we get $-9x^2 - 15x - 11.$

Example 2. Expand $\dfrac{4x^2y + 10x^3y^{-1} + 12xy^{-3}}{30xy^2}$ into a sum of three separate terms.

Solution. In this expression, each of the terms in the numerator is divided by $30xy^2$; that is, the expression is equivalent to

$$\frac{1}{30xy^2} (4x^2y + 10x^3y^{-1} + 21xy^{-3}).$$

Expanding by multiplication we get

$$\frac{4x^2y}{30xy^2} + \frac{10x^3y^{-1}}{30xy^2} + \frac{12xy^{-3}}{30xy^2} \,,$$

or

$$\frac{2x}{15y} + \frac{x^2}{3y^3} + \frac{2}{5y^5} \,.$$

Certain special products occur so often that it is worthwhile to memorize them:

(1) $(a + b)^2 = a^2 + 2ab + b^2$ (not $a^2 + b^2$)

(2) $(a - b)^2 = a^2 - 2ab + b^2$

(3) $(a + b)(a - b) = a^2 - b^2$

(4) $(a + b)^3 = a^3 + 3a^2b + 3ab^2 + b^3$

(5) $(a - b)^3 = a^3 - 3a^2b + 3ab^2 - b^3$

(6) $(a + b + c)^2 = a^2 + b^2 + c^2 + 2ab + 2ac + 2bc$

(7) The Binomial Theorem: If n is a positive integer, $(a + b)^n =$
$a^n + na^{n-1}b + \dfrac{n(n - 1)}{2}\, a^{n-2}b^2 + \dfrac{n(n - 1)(n - 2)}{3 \cdot 2 \cdot 1}\, a^{n-3}b^3 + \ldots + b^n.$

Example 3. Expand $(2x + 3y - 4)^2$.

Solution. We can apply the identity in (6) directly, with a = 2x, b = 3y, and

c = -4, to obtain
$$4x^2 + 9y^2 + 16 + 12xy - 16x - 24y.$$

Another way to carry out this multiplication is to arrange the subproducts in the familiar long form:

$$2x + 3y - 4$$
$$\underline{2x + 3y - 4}$$
$$4x^2 + 6xy - 8x$$
$$6xy \qquad + 9y^2 - 12y$$
$$\underline{\qquad\qquad - 8x \qquad\quad - 12y + 16}$$
$$4x^2 + 12xy - 16x + 9y^2 - 24y + 16$$

43

EXERCISES

Expand the following expressions by multiplication and simplify.

1. $(2x + 1)(3x + 5)$

2. $(-3x + 7)(2x - 6)$

3. $(8x + 3y)^2$

4. $(2x + 3)(2x - 3)$

5. $(x + 2)^3$

6. $\dfrac{60x^3 + 24x^2 - 36x}{12x}$

7. $\dfrac{25a^4b^3 - 35a^7b^4}{5a^2b^2}$

8. $(3x + 1)(7x - 2) - (3x + 4)(6x - 1) - x(2x - 19)$

9. $(8x - 7y)(8x + 7y) + (x - y)^2 - (2x + 4y)(2x - 4y)$

10. $(x^3 + 3x^2 - 2x + 3)(x^2 + 5x - 2)$

11. $(x - 2y)^3$

ANSWERS

1. $6x^2 + 13x + 5$ 2. $-6x^2 + 32x - 42$ 3. $64x^2 + 48xy + 9y^2$

4. $4x^2 - 9$ 5. $x^3 + 6x^2 + 12x + 3$ 6. $5x^2 + 2x - 3$

7. $5a^2b - 7a^5b^2$ 8. $x^2 - x + 2$ 9. $61x^2 - 2xy - 32y^2$

10. $x^5 + 8x^4 + 11x^3 - 13x^2 + 19x - 6$ 11. $x^3 - 6x^2y + 12xy^2 - 8y^3$

12.   Factoring

In a sense, factoring simply reverses the operation of expansion discussed in the preceding section. For example, each of the terms in the expression $-4x^4 - 6x^3 - 2x^2$ has a common factor of $-2x^2$ which may be factored out; that is,

$$-4x^4 - 6x^3 - 2x^2 = -2x^2(2x^2 + 3x + 1).$$

44

Furthermore, you can check that $2x^2 + 3x + 1$ is the product of $(2x + 1)$ and $(x + 1)$. Thus, $-4x^2 - 6x^3 - 2x^2$ in factored form is

$$-2x^2(2x + 1)(x + 1).$$

Factoring polynomial expressions is important in simplifying algebraic expressions and, as you will see in later sections, in solving polynomial equations.

Example 1. Factor the numerator and denominator and simplify

$$\frac{3x^3 + 3x^2 - 18x}{4x^2 - 3x - 10}.$$

Solution. We first consider the numerator, looking for factors common to all the terms. In this case, we can factor out $3x$ to get $3x(x^2 + x - 6)$. If $x^2 + x - 6$ can be factored, it must have the form

$$(x + \underline{\ ?\ })(x + \underline{\ ?\ }),$$

where the two terms belonging in the blanks must have a product of -6 and a sum of 1. The possibilities (in integers) are given by the pairs $(6,-1)$, $(1,-6)$, $(2,-3)$, and $(3,-2)$; by trial and error, we find that $(3,-2)$ works. Thus the numerator can be factored into the form $3x(x + 3)(x - 2)$. The denominator, if factorable, must have one of the forms

$$(4x \pm \underline{\quad})(x \pm \underline{\quad}) \quad \text{or} \quad (2x \pm \underline{\quad})(2x \pm \underline{\quad}),$$

and the blanks must be filled with the pairs $(5,2)$, $(2,5)$, $(1,10)$, or $(10,1)$. Again, by trial and error, we find the factorization $(4x + 5)(x - 2)$. Therefore,

$$\frac{3x^3 + 3x^2 - 18x}{4x^2 - 3x - 10} = \frac{3x(x + 3)(x - 2)}{(4x + 5)(x - 2)} = \frac{3x(x + 3)}{4x + 5}.$$

Factoring trinomials of the form $ax^2 + bx + c$ takes practice, but you will soon learn to discard certain combinations quickly to abbreviate the work.

Example 2. Simplify

$$(x + 1)(x + 4)^2(2x + 5) + (x + 1)^2(x + 4)(2x + 5) + (x + 1)^2(x + 4)^2.$$

Solution. We see that $(x + 1)$ and $(x + 4)$ appear as factors in each of the terms, so we can write the sum in the form

$$(x + 1)(x + 4)[(x + 4)(2x + 5) + (x + 1)(2x + 5) + (x + 1)(x + 4)],$$

$$(x + 1)(x + 4)[(2x^2 + 13x + 20) + (2x^2 + 7x + 5) + (x^2 + 5x + 4)],$$

$$(x + 1)(x + 4)(5x^2 + 25x + 29),$$

and the final quadratic term doesn't factor further (without introducing radicals).

Expressions in certain forms can always be factored immediately. These would include the following:

Difference of Squares: $\quad a^2 - b^2 = (a - b)(a + b)$

Difference of Cubes: $\quad a^3 - b^3 = (a - b)(a^2 + ab + b^2)$

Sum of Cubes: $\quad a^3 + b^3 = (a + b)(a^2 - ab + b^2)$

Difference of n-th Powers (n a positive integer):

$$a^n - b^n = (a - b)(a^{n-1} + a^{n-2}b + a^{n-3}b^2 + \ldots + ab^{n-2} + b^{n-1})$$

Example 3. Simplify the product

$$\frac{9x^2 - 6x + 4}{25 - 4x^2} \cdot \frac{6x^2 + 19x + 10}{27x^3 + 8}.$$

Solution. $25 - 4x^2$ is a difference of squares, so is equal to $(5 - 2x)(5 + 2x)$; $27x^3 + 8$ is a sum of cubes, so is equal to $(3x + 2)(9x^2 - 6x + 4)$; $6x^2 + 19x + 10$ factors into $(2x + 5)(3x + 2)$. Therefore, the given product is

46

equal to

$$\frac{(9x^2 - 6x + 4)}{(5 - 2x)(5 + 2x)} \cdot \frac{(2x + 5)(3x + 2)}{(3x + 2)(9x^2 - 6x + 4)} \cdot$$

By cancelling we get

$$\frac{1}{5 - 2x} \cdot$$

EXERCISES

Simplify the following expressions by factoring.

1. $15ab^2 - 20a^2b^3 - 25a^5b^7$

2. $2(x + 1)(x - 3)^3 + 3(x + 1)^2(x - 3)^2$

3. $3(x + 1)^2(x + 4)^3 + 3(x + 1)^3(x + 4)^2$

4. $x^2 + 8x + 15$

5. $4x^2 - 13x - 12$

6. $\dfrac{25x^2y^4 - 9x^6}{125x^3y^6 + 27x^9}$

7. $\dfrac{x^2 + 6x - 16}{x^2 - 9x + 14} \cdot \dfrac{x^2 - 8x + 7}{x^2 + 10x + 16}$

8. $\dfrac{15a^4b^9}{18 - 9x - 2x^2} \cdot \dfrac{3x^2 + 16x - 12}{25a^9b^3} \cdot \dfrac{40a^2x - 60a^2}{3bx - 2b}$

9. $\dfrac{2y(y + 1) - y^2}{(y + 1)^2}$

10. $\dfrac{4x(x^2 + 1)(x + 2)^3 - 3(x^2 + 1)^2(x + 2)^2}{(x + 2)^6}$

47

1. $5ab^2(3 - 4ab - 5a^4b^5)$ 2. $(x + 1)(x - 3)^2(5x - 3)$

3. $3(x + 1)^2(x + 4)^2(2x + 5)$ 4. $(x + 5)(x + 3)$ 5. $(4x + 3)(x - 4)$

6. $\dfrac{5y^2 - 3x^2}{25xy^4 - 15x^3y^2 + 9x^5}$ 7. $\dfrac{x - 1}{x + 2}$ 8. $\dfrac{-12b^5}{a^3}$ 9. $\dfrac{y(y + 2)}{(y + 1)^2}$

10. $\dfrac{(x^2 + 1)(x^2 + 8x - 3)}{(x + 2)^4}$

---

13. Square Roots, Rationalization

If a is non-negative, $\sqrt{a}$ will denote the non-negative square root of a (we will write $\pm\sqrt{a}$ if we wish to consider both square roots of a, and $-\sqrt{a}$ for the negative square root).

The main multiplicative properties of square roots are the following:

$$\sqrt{a}\sqrt{a} = a$$

$$\sqrt{a}\sqrt{b} = \sqrt{ab}$$

$$\frac{\sqrt{a}}{\sqrt{b}} = \sqrt{\frac{a}{b}}$$

These rules hold without exception for positive numbers, but care must be observed if they are applied to other numbers. For that reason we restrict attention in this section to positive numbers. With these rules, it is often possible to simplify complicated quantities containing square roots.

<u>Example 1.</u> Add and simplify

$$\sqrt{75x^7y} + \sqrt{27x^5y^3} - \sqrt{12x^3y^5}.$$

<u>Solution.</u> We first simplify each of the three terms in the sum by factoring out perfect squares under the radical sign. Thus,

$$\sqrt{75x^7y} = \sqrt{5^2 \cdot 3 \cdot (x^3)^2 xy} = 5x^3\sqrt{3xy},$$

$$\sqrt{27x^5y^3} = 3x^2y\sqrt{3xy},$$

$$\sqrt{12x^3y^5} = 2xy^2\sqrt{3xy}.$$

Therefore the given sum is equal to

$$5x^3\sqrt{3xy} + 3x^2y\sqrt{3xy} - 2xy^2\sqrt{3xy},$$

from which the common factor $x\sqrt{3xy}$ may be taken out to give

$$x(5x^2 + 3xy - 2y^2)\sqrt{3xy}.$$

In this case the expression within the parentheses further factors to give a final expression

$$x(5x - 2y)(x + y)\sqrt{3xy}.$$

<u>Example 2.</u> Expand and simplify $(2\sqrt{3x} - 4\sqrt{2x})(3\sqrt{3x} + \sqrt{2x})$.

<u>Solution.</u> Multiplying in the usual way we get

$$6(3x) - 10\sqrt{2x}\sqrt{3x} - 4(2x)$$

which is equal to

$$10x - 10x\sqrt{6}.$$

Factoring out 10x we arrive at the expression

$$10x(1 - \sqrt{6}), \quad \text{or equivalently,} \quad -10(\sqrt{6} - 1)x.$$

Example 3. Add and simplify $\sqrt{\dfrac{1}{2x}} - \sqrt{\dfrac{1}{2x^3}} - \sqrt{\dfrac{4}{2x^5}}$.

Solution. Proceeding as in Example 1, we get

$$\sqrt{\dfrac{1}{2x}} - \dfrac{1}{x}\sqrt{\dfrac{1}{2x}} - \dfrac{2}{x^2}\sqrt{\dfrac{1}{2x}} ,$$

or

$$(1 - \dfrac{1}{x} - \dfrac{2}{x^2})\sqrt{\dfrac{1}{2x}} .$$

The expression within the parentheses can be added by taking x^2 as a common

denominator

$$(\dfrac{x^2 - x - 2}{x^2})\sqrt{\dfrac{1}{2x}} ,$$

and after factoring the numerator we have

$$\dfrac{(x - 2)(x + 1)}{x^2 \cdot \sqrt{2x}} .$$

We may eliminate the square root in the denominator of this expression by mul-

tiplying the numerator and denominator by $\sqrt{2x}$, to get

$$\dfrac{(x - 2)(x + 1)\sqrt{2x}}{2x^3} .$$

Example 4. Express the quantity

$$\dfrac{2\sqrt{3} - \sqrt{2}}{\sqrt{3} + 2\sqrt{2}}$$

in an equivalent form containing no square roots in the (a) denominator, (b)

numerator.

Solution. The procedure for doing this (called *rationalizing the denominator*

(numerator)) consists of multiplying the numerator and denominator by the

appropriate *conjugate*. The trick is to notice that when $\sqrt{a} + \sqrt{b}$ is multiplied

by the conjugate $\sqrt{a} - \sqrt{b}$, the result is a - b which is free of square roots.

(a) $\dfrac{2\sqrt{3} - \sqrt{2}}{\sqrt{3} + 2\sqrt{2}} = \dfrac{2\sqrt{3} - \sqrt{2}}{\sqrt{3} + 2\sqrt{2}} \cdot \dfrac{\sqrt{3} - 2\sqrt{2}}{\sqrt{3} - 2\sqrt{2}} = \dfrac{2 \cdot 3 - 5\sqrt{6} + 2 \cdot 2}{3 - 4 \cdot 2}$

$\qquad\qquad = \dfrac{10 - 5\sqrt{6}}{-5} = -2 + \sqrt{6} = -(2 - \sqrt{6}).$

(b) $\dfrac{2\sqrt{3} - \sqrt{2}}{\sqrt{3} + 2\sqrt{2}} = \dfrac{2\sqrt{3} - \sqrt{2}}{\sqrt{3} + 2\sqrt{2}} \cdot \dfrac{2\sqrt{3} + \sqrt{2}}{2\sqrt{3} + \sqrt{2}} = \dfrac{4 \cdot 3 - 2}{2 \cdot 3 + 5\sqrt{6} + 2 \cdot 2}$

$\qquad\qquad = \dfrac{10}{10 + 5\sqrt{6}} = \dfrac{2}{2 + \sqrt{6}} \; .$

<u>Example 5.</u> Add and simplify $2x\sqrt{x^2 + 1} + \dfrac{x^3}{\sqrt{x^2 + 1}} \; .$

<u>Solution.</u> Combining the terms with a common denominator, we get

$$\dfrac{2x(x^2 + 1) + x^3}{\sqrt{x^2 + 1}} \; ,$$

or

$$\dfrac{x(3x^2 + 2)}{\sqrt{x^2 + 1}} \; .$$

In working with square roots, note that:

$$\sqrt{3x} \text{ is NOT equal to } 3\sqrt{x},$$

$$\sqrt{a} + \sqrt{b} \text{ is NOT equal to } \sqrt{a + b},$$

$$\sqrt{a^2 + b^2} \text{ is NOT equal to } a + b,$$

$$\sqrt{1 + \dfrac{x^2}{y^2}} \text{ is NOT equal to } \sqrt{y^2 + x^2}.$$

These may be checked out by considering special cases; for example, $\sqrt{9 + 16} = \sqrt{25} = 5$, whereas $\sqrt{9} + \sqrt{16} = 3 + 4 = 7.$

Our decision to regard $\sqrt{x}$ as the non-negative square root of x has some subtle implications that are often overlooked. Notice, for example, that $\sqrt{(-2)^2} = \sqrt{4} = 2$ (rather than -2). That is to say, it may not be true that $\sqrt{x^2} = x$. The fact is that $\sqrt{x^2} = x$ only if x is non-negative; otherwise $\sqrt{x^2}$ equals -x. This is expressed more succinctly by the equation

$$\sqrt{x^2} = |x|$$

which is valid for every real number x.

It is most convenient when learning the computational aspects of manipulating square roots, to assume that all letters under the square roots take on only positive values. Thus, in Example 1, when we found

$$\sqrt{75x^7y} = 5x^3\sqrt{3xy}$$

we tacitly assumed that x and y were both non-negative. If, however, both x and y are negative, then $\sqrt{75x^7y}$ makes sense since $75x^7y$ is positive, but

$$\sqrt{75x^7y} = -5x^3\sqrt{3xy} = 5|x|^3\sqrt{3xy} .$$

There are times in the calculus when this kind of analysis is necessary.

EXERCISES

In the following exercises, assume that all letters represent positive quantities.

Perform the computations below and simplify.

1. $\sqrt{5a^5} - \sqrt{245a^3} - \sqrt{320a}$

2. $5\sqrt{\dfrac{1}{8x}} - 2\sqrt{\dfrac{1}{18x}} + 3\sqrt{\dfrac{1}{50x}}$

3. $\sqrt{10a^3b}\ \sqrt{8ab^2}\ \sqrt{5a^2b^3}$

4. $\sqrt{2}(\sqrt{32} - \sqrt{3}) - \sqrt{3}(\sqrt{2} + \sqrt{12})$

5. $(\sqrt{x} + 4\sqrt{y})(2\sqrt{x} - 3\sqrt{y})$

6. $\dfrac{1/a}{\sqrt{1 - (x/a)^2}}$

7. $(2x + 2)\sqrt{x + 1} - \dfrac{x^2 + 2x + 3}{2\sqrt{x + 1}}$

Find an equivalent expression without square roots in the (a) denominator,
(b) numerator.

8. $\dfrac{\sqrt{2} - \sqrt{3}}{\sqrt{2} + \sqrt{3}}$

9. $\dfrac{2\sqrt{3} - 4\sqrt{5}}{\sqrt{3} + \sqrt{5}}$

10. $\dfrac{\sqrt{x + h} - \sqrt{x}}{\sqrt{x + h} + \sqrt{x}}$

ANSWERS

1. $(a - 8)(a + 1)\sqrt{5a}$ 2. $\dfrac{73\sqrt{2x}}{60x}$ 3. $20a^3b^3$ 4. $2(1 - \sqrt{6})$

5. $2x + 5\sqrt{xy} - 12y$ 6. $\dfrac{1}{\sqrt{a^2 - x^2}}$ 7. $\dfrac{3x^2 + 6x + 1}{2\sqrt{x + 1}}$

8. $2\sqrt{6} - 5$, $\dfrac{-1}{5 + 2\sqrt{6}}$ 9. $3\sqrt{15} - 13$, $\dfrac{-34}{13 + 3\sqrt{15}}$

10. $\dfrac{2x + h - 2\sqrt{x(x + h)}}{h}$, $\dfrac{h}{2x + h + 2\sqrt{x(x + h)}}$

---

14.   Radicals, Rational Exponents

If n is a positive integer, an nth root of a number a is any number x such that $x^n = a$. The *principal nth root* of a is the term used to denote the one real root of a if n is odd and the one non-negative real root of a if a is non-negative and n is even. We denote the principal nth root of a by $\sqrt[n]{a}$. For example,

$$\sqrt[3]{8} = 2,$$

$$\sqrt[4]{16} = 2 \quad (\text{not} \pm 2),$$

$$\sqrt[3]{-27} = -3.$$

The multiplicative properties of nth roots are similar to those of square roots, namely

$$\sqrt[n]{a^n} = a,$$

$$\sqrt[n]{ab} = \sqrt[n]{a}\,\sqrt[n]{b},$$

$$\sqrt[n]{\frac{a}{b}} = \frac{\sqrt[n]{a}}{\sqrt[n]{b}}.$$

When n = 2 we have the square root properties discussed in Section 12. As in that section, these rules hold without exception when the values of a and b are positive; otherwise, they are not necessarily correct, and care must be taken. We therefore assume throughout this section that the letters contained in the radicals represent positive quantities.

Example 1. Simplify each of the following quantities:

a) $\sqrt[3]{72a^7b^9}$

b) $\sqrt[4]{\dfrac{32x^7}{125y^9}}$

Solution. For (a), factor out "perfect cubes":

$$\sqrt[3]{72a^7b^9} = \sqrt[3]{2^3 \cdot 3^2 \cdot a^6 \cdot a \cdot b^9} = 2a^2b^3\sqrt[3]{9a}.$$

Similarly, with (b) we have

$$\sqrt[4]{\frac{32x^7}{125y^9}} = \sqrt[4]{\frac{2^4 \cdot 2 \cdot x^4 \cdot x^3}{5^3 \cdot y^8 \cdot y}} = \frac{2x}{y^2}\sqrt[4]{\frac{2x^3}{5^3y}}$$

It is often desirable to eliminate radicals in the denominator; this can be done by multiplying numerator and denominator by the same appropriate quantity. In this case, we notice that 5^3y becomes a perfect fourth power when it is multiplied by $5y^3$. Therefore, we have

$$\frac{2x}{y^2} \sqrt[4]{\frac{2x^3}{5^3y}} = \frac{2x}{y^2} \sqrt[4]{\frac{2x^3}{5^3y} \cdot \frac{5y^3}{5y^3}} = \frac{2x}{y^2} \sqrt[4]{\frac{10x^3y^3}{5^4y^4}}$$

$$= \frac{2x}{5y^3} \sqrt[4]{10x^3y^3}.$$

Computations involving radicals are simplified by the use of *fractional powers*. More specifically, we extend the definition of a^r given in Section 10 so that it has a meaning when r is any fraction (rational number). If n is a positive integer, we define

$$a^{1/n} = \sqrt[n]{a} \ ,$$

and if $\frac{m}{n}$ is any rational number with n a positive integer, we define

$$a^{m/n} = \sqrt[n]{a^m} \ ,$$

or, using fractional exponents,

$$a^{m/n} = (a^m)^{1/n}.$$

It is remarkable that this definition makes sense and extends the usual rules of exponents (see Section 10) so that they continue to hold for fractional powers. In particular, if $\frac{m}{n}$ is a rational number with n a positive integer,

$$a^{m/n} = (a^{1/n})^m, \quad \text{or equivalently,} \quad a^{m/n} = (\sqrt[n]{a})^m,$$

and, if s and t are rational numbers and a and b are positive reals,

$$(1) \quad a^s a^t = a^{s+t},$$

$$(2) \quad (a^s)^t = a^{st},$$

$$(3) \quad (ab)^s = a^s b^s$$

$$(4) \quad \frac{a^s}{a^t} = a^{s-t} = \frac{1}{a^{t-s}} \, ,$$

$$(5) \quad \left(\frac{a}{b}\right)^s = \frac{a^s}{b^s} \, .$$

__Example 2.__ Evaluate a) $(36)^{3/2}$,   b) $(-32)^{-4/5}$.

__Solution.__ a) $(36)^{3/2} = (6^2)^{3/2} = 6^3 = 216.$

b) $(-32)^{-4/5} = [(-2)^5]^{-4/5} = (-2)^{-4} = 1/16.$

The rules of exponents are useful in simplifying terms containing many higher order radicals. The procedure is shown in Example 3.

__Example 3.__ Express each of the following quantities in terms of a single radical.

a) $\sqrt[3]{\sqrt{5}}$,   b) $\sqrt[4]{25 \sqrt[3]{5}}$,   c) $\sqrt[3]{x} \, \sqrt[4]{x^3}$.

__Solution.__ In each case, transform into exponential form and apply the rules of exponents.

a) $\sqrt[3]{\sqrt{5}} = (5^{1/2})^{1/3} = 5^{1/6} = \sqrt[6]{5}$.

b) $\sqrt[4]{25 \sqrt[3]{5}} = (5^2 \cdot 5^{1/3})^{1/4} = (5^{7/3})^{1/4} = 5^{7/12} = \sqrt[12]{5^7}$.

c) $\sqrt[3]{x} \, \sqrt[4]{x^3} = (x)^{1/3}(x^3)^{1/4} = x^{1/3} x^{3/4} = x^{\frac{1}{3}+\frac{3}{4}} = x^{13/12}$

$= x \sqrt[12]{x}$.

__Example 4.__ Add and simplify $3(x^2 + 1)^{2/3} + 4x^2(x^2 + 1)^{-1/3}$.

__Solution.__ As in working with integer exponents, we first eliminate negative exponents,

56

$$3(x^2 + 1)^{2/3} + \frac{4x^2}{(x^2 + 1)^{1/3}} \quad ,$$

and then combine after finding a common denominator:

$$\frac{3(x^2 + 1) + 4x^2}{(x^2 + 1)^{1/3}} \quad , \quad \text{or equivalently,} \quad \frac{7x^2 + 3}{(x^2 + 1)^{1/3}}$$

EXERCISES

1. Evaluate the following quantities. a) $16^{3/2}$ b) $(.008)^{1/3}$

   c) $\sqrt{-25(\sqrt[3]{-64}\,)}$

2. Solve the following equation for z when x = 4 and y = 4.

   $$\frac{4}{3}\,(2x)^{-2/3}\,y^{1/2} + \frac{1}{2}\,(2x)^{1/3}\,y^{-1/2}\,z \;=\; 0.$$

3. Express each of the following quantities using rational exponents instead of radical signs. a) $\sqrt[4]{x^3}$ b) $\sqrt{\sqrt{x}}$ c) $x\,\sqrt[3]{x}$

Simplify each of the following expressions.

4. a) $(x^{2/3})(x^{1/3})^4$ b) $\dfrac{x^{-3/4}}{x^{3/2}}$

5. a) $\dfrac{(x^{1/3}\cdot y^{3/4})^2}{(x^{2/3}\cdot y^{1/2})^3}$ b) $\left(\dfrac{y^{16}}{81x^{-12}}\right)^{1/4}\left(\dfrac{-x^9}{27y^{-15}}\right)^{-1/3}$

6. Express the following expressions in terms of a single radical.

   a) $\sqrt{\sqrt[3]{x^5}}$ b) $\sqrt[3]{x^2}\;\sqrt[2]{x}$

7. If $y = \dfrac{V}{\pi x^2}$, show that $y = x$ when $x = \sqrt[3]{\dfrac{V}{\pi}}$. (Substitute $x = \sqrt[3]{\dfrac{V}{\pi}}$ into

   the equation $y = \dfrac{V}{\pi x^2}$.)

Add and simplify each of the following expressions.

8. $3(x^2 + 1)^{2/3} + 4x(x^2 + 1)^{-1/3}$

9. $2(x + 1)^{-1/3}(x^2 + 1)^{1/2} + 3x(x + 1)^{2/3}(x^2 + 1)^{-1/2}$

10. By factoring $\sqrt{x + 1}$ from both terms, simplify

$$4x(x + 1)^{3/2} + 3(x^2 + 2)(x + 1)^{1/2}.$$

11. Eliminate the radicals in the numerator (that is, rationalize the numerator) of

$$\frac{(x + h)^{1/3} - x^{1/3}}{h}$$

by multiplying and dividing by

$$(x + h)^{2/3} + (x + h)^{1/3}x^{1/3} + x^{2/3}.$$

(This complicated choice is not so mysterious if you recall that $a^3 - b^3 = (a - b)(a^2 + ab + b^2)$ and consider the special case $a = (x + h)^{1/3}$ and $b = x^{1/3}$.)

ANSWERS

1. a) 64, b) .2, c) 10   2. $-4/3$ 3. a) $x^{3/4}$, b) $x^{1/4}$, c) $x^{4/3}$

4. a) x^2, b) $x^{-9/4}$ 5. a) $x^{-4/3}$, b) $-1/y$ 6. a) $\sqrt[6]{x^5}$, b) $x\sqrt[6]{x}$

8. $\dfrac{3x^2 + 4x + 3}{(x^2 + 1)^{1/3}}$ 9. $\dfrac{5x^2 + 3x + 2}{(x + 1)^{1/3}(x^2 + 1)^{1/2}}$ 10. $\sqrt{x + 1}\ [7x^2 + 4x + 6]$

11. $\dfrac{1}{(x + h)^{2/3} + (x + h)^{1/3}x^{1/3} + x^{2/3}}$

15. Functions, Notation

A basic mathematical idea which is central to the understanding of calculus is the concept of a function. In this section we will discuss the notational aspects of functions, since this often is the source of confusion among students. For a more complete understanding of why functions are important and how they arise in applications, you should carefully study Section 23.

A *function* f defined on a certain set of real numbers D is a rule which assigns a real number to each element of D. If x is an element of D, the number which is assigned to x is called the *value* of the function f at x and is denoted by f(x) (this is read "f at x"). The set D is called the *domain* of F.

Consider the function f which assigns to each real number x the real number 2x + 1. This rule is expressed more compactly by the formula

$$f(x) = 2x + 1.$$

The parentheses in this notation do not mean f times x; rather, the formula gives the value of the quantity that is assigned by f to x (or, to the quantity within the parentheses). To find what number is assigned to 3 by this function, simply substitute 3 for x and find that

$$f(3) = 2 \cdot 3 + 1 = 7,$$

that is, the function assigns 7 to 3. Similarly,

$$f(-1) = 2(-1) + 1 = -1,$$
$$f(0) = 2 \cdot 0 + 1 = 1,$$
$$f(1) = 2 \cdot 1 + 1 = 3,$$

and so forth.

Note that this function could be described as well by any of the formulas

$$f(s) = 2s + 1, \quad f(t) = 2t + 1, \quad f(w) = 2w + 1.$$

That is to say, the *independent variable* in the description of the function may be different from x. However, variables are generally denoted by lower case letters such as u, x, y, z. Similarly, other letters may be used instead of f to denote functions; capital letters are frequently used.

It is extremely helpful and essential in calculus to think of functions geometrically. Suppose that f is a function defined on the interval [a,b] ([a,b] denotes the set of numbers x for which a ≤ x ≤ b). This should conjure up an image of a graph such as the one shown:

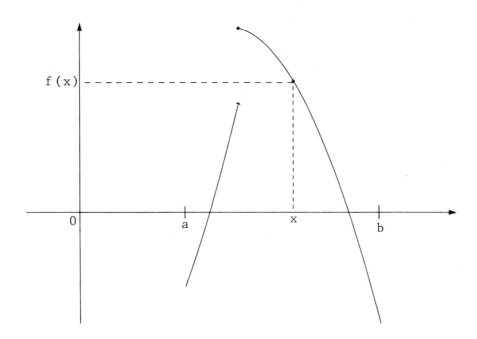

Such a curve, called the graph of f, consists of the points (x,f(x)) as x varies through the interval [a,b]. You should think of the graph of f dynamically: each point x along the horizontal axis between a and b corresponds to a point on the vertical line through x.

The graph of f(x) = 2x + 1 is the straight line of slope 2 and y-intercept 1.

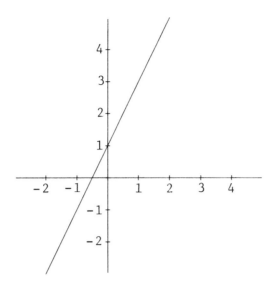

Example 1. A function f is defined by $f(x) = \sqrt{x}$.

a) Sketch the graph of f.

b) Compute $f(a)$, $f(2x)$, and $f(x + h)$.

Solution. Although the domain of f is not explicitly given, it is understood that the domain is the set of non-negative numbers (because these are the only values for which the square root is defined as a real number).

As in Section 7, we can sketch the graph of f after making a table and plotting a few points.

x	f(x)
0	0
1	1
4	2
9	3

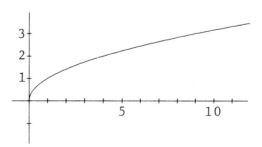

For part (b): $f(a) = \sqrt{a}$; $f(2x) = \sqrt{2x}$ (remove x in the definition of f and replace it with 2x); $f(x + h) = \sqrt{x + h}$ (replace x in the definition of

f with x + h).

Note that the rule F given by $F(x) = \pm\sqrt{x}$ does NOT define a function.   The

reason is that F assigns more than one number to x (when x is positive), but

to be a function a rule must assign a *single value* to each element in the do-

main.   Similarly, neither of the graphs

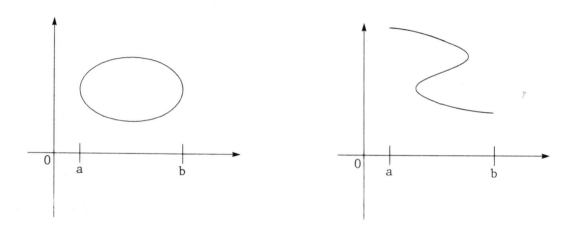

represent functions over the interval [a,b].

Example 2.   A function f is defined by $f(x) = \dfrac{1}{1 + x}$.

   a) Sketch the graph of f.

   b) Compute $f(2x)$; $f(x + h)$; $f(\frac{1}{x})$.

Solution.   a) The domain of f is the set of all real numbers except x = -1

(when x = -1, f is undefined since division by 0 is not defined).   As x takes

on values close to -1, the denominator 1 + x is small which in turn means that

$\dfrac{1}{1 + x}$ is large in magnitude: a large positive number if x is slightly larger

than -1 and a large negative number if x is slightly smaller than -1.

62

x	f(x)
2	1/3
1	1/2
0	1
-1/2	2
-2/3	3
-4/3	-3
-3/2	-2
-2	-1
-3	-1/2
-4	-1/3

b) $f(2x) = \dfrac{1}{1 + 2x}$; $f(x + h) = \dfrac{1}{1 + (x + h)}$; $f\left(\dfrac{1}{x}\right) = \dfrac{1}{1 + 1/x} = \dfrac{x}{x + 1}$.

Example 3. Let $f(x) = 3x + 2$, $g(x) = -x + 4$. (f and g are defined for all real numbers.) Compute $f(2x + 1)$, $g(2/(x^2 + 3))$, $f(g(x))$, $g(f(x))$.

Solution. The function f is defined by

$$f(\boxed{}) = 3(\boxed{}) + 2\,,$$

where whatever is between the parentheses on the left is placed between the parentheses on the right. Thus

$$f(\boxed{2x + 1}) = 3(\boxed{2x + 1}) + 2$$

$$= 6x + 5.$$

The function g is defined by

$$g(\boxed{}) = -(\boxed{}) + 4,$$

so by direct substitution

$$g\left(\left[\frac{2}{x^2 + 3}\right]\right) = -\left(\left[\frac{2}{x^2 + 3}\right]\right) + 4$$

$$= \frac{-2 + 4(x^2 + 3)}{x^2 + 3}$$

$$= \frac{4x^2 + 10}{x^2 + 3} \; .$$

Similarly, $f(g(x)) = f(-x + 4) = 3(-x + 4) + 2 = -3x + 14$, or equivalently,

$$f(g(x)) = 3(g(x)) + 2 = 3(-x + 4) + 2 = -3x + 14;$$

$$g(f(x)) = g(3x + 2) = -(3x + 2) + 4 = -3x + 2,$$

or equivalently,

$$g(f(x)) = -(f(x)) + 4 = -(3x + 2) + 4 = -3x + 2.$$

EXERCISES

1. What is the domain of the function f defined by $f(x) = \sqrt{x - 2}$?

2. What is the domain of the function g defined by $g(x) = \dfrac{1}{(x - 1)(x + 2)}$?

3. If $f(x) = 3x + 2$, find $f(2)$, $f(-2)$, $f(a)$.

4. If $f(x) = \dfrac{1}{x} + x$, find $f(1)$, $f(2x)$, $f(x + 1)$.

5. If $f(x) = -x + 3$, find $f(-2)$, $f(x + 3)$, $f(x^2)$.

6. A function g is defined by $g(x) = x^2$. a) Sketch the graph of g. b) Compute $g(x - 1)$; $g(x + h)$.

7. If $f(x) = \sqrt{x^2 + 9}$, find $f(4) - f(0)$.

8. If $f(x) = \frac{2}{3}(x + 1)^{3/2}$, find $f(3) - f(0)$.

9. If $f(x) = x + 2$, $g(x) = -2x + 1$, compute $f(g(x))$ and $g(f(x))$.

10. If $f(x) = \frac{x^2}{(3x + 2)^{2/3}}$, show that $f(\frac{y-2}{3}) = \frac{1}{9}[y^{4/3} - 4y^{1/3} + 4y^{-2/3}]$.

ANSWERS

1. $x \geq 2$ 2. All real numbers except 1 and -2   3. $f(2) = 8$, $f(-2) = -4$,

$f(a) = 3a + 2$ 4. $f(1) = 2$, $f(2x) = 1/2x + 2x$, $f(x + 1) = \frac{1}{x + 1} + x + 1$

5. $f(-2) = 5$, $f(x + 3) = -x$, $f(x^2) = -x^2 + 3$ 6. a) The graph of g is

the same as that given in Example 1, Section 7, page 19.

b) $g(x - 1) = (x - 1)^2$, $g(x + h) = (x + h)^2$ 7. 2   8. 14/3

9. $f(g(x)) = -2x + 3$, $g(f(x)) = -2x - 3$

16. Solving Equations, Complex Fractions

The methods for working the exercises in this section have been discussed in the preceding sections (particularly in Sections 5 and 6). The problems in this section are slightly more difficult; however, they are more representative of the kind of computation you will encounter as you study the derivative and its applications in calculus.

Example 1. Solve for y' in the following equation:

$$6(x^2 + y^2)(2x + 2yy') = 25(2x - 2yy').$$

Solution. We begin by expanding (since there are no negative exponents and no fractions):

$$12x(x^2 + y^2) + 12y(x^2 + y^2)y' = 50x - 50yy'.$$

Then, separating variables we get

$$[12y(x^2 + y^2) + 50y]y' = 50x - 12x(x^2 + y^2),$$

$$y' = \frac{50x - 12x(x^2 + y^2)}{12y(x^2 + y^2) + 50y}.$$

The remaining step is to simplify this last expression by factoring the numerator and denominator as much as possible:

$$y' = \frac{2x[25 - 6x^2 - 6y^2]}{2y[6x^2 + 6y^2 + 25]} = -\frac{x(6x^2 + 6y^2 - 25)}{y(6x^2 + 6y^2 + 25)}.$$

Example 2. Solve for x: $\dfrac{2x^2}{x^2 - 3x - 10} - \dfrac{8}{x - 5} + \dfrac{1}{x + 2} = 2.$

Solution. We begin by clearing the fractions. Since $x^2 - 3x - 10 = (x - 5)(x + 2)$, we multiply each side of the equation by $(x - 5)(x + 2)$ to obtain

$$2x^2 - 8(x + 2) + (x - 5) = 2x^2 - 6x - 20,$$

$$- x = 1,$$

$$x = -1.$$

This value for x satisfies the original equation.

Note: When an equation is multiplied by an expression containing the variable, that multiplication may introduce *extraneous* roots. For example, when $(x^2 - 4)/(x - 2) = 0$ is multiplied by $x - 2$, we get $x^2 - 4 = 0$ which has two solutions, $x = 2$ and $x = -2$. The solution $x = -2$ is a solution to the original equation, but $x = 2$ is not because $0/0$ is meaningless. We say that $x = 2$ is an extraneous root. The general rule is this: suppose that an equation can be put into the form $\dfrac{P(x)}{Q(x)} = 0$, where $P(x)$ and $Q(x)$ are polynomials

with no common factors. If this equation is now multiplied by $Q(x)$, the solutions of the original equation are those of $P(x) = 0$ that are not solutions of $Q(x) = 0$.

Example 3. Simplify
$$\frac{2 - \dfrac{x - 8}{x^2 - 2x - 3}}{\dfrac{12x + 1}{x + 3} - \dfrac{6x - 17}{x - 3}} \; .$$

Solution. A fraction that has simple fractions in the numerator or denominator is called a complex fraction. The procedure, as noted in previous sections, is to simplify the numerator and denominator separately, and then to invert the denominator and multiply. In this case, the numerator is

$$2 - \frac{x - 8}{x^2 - 2x - 3} = \frac{2(x^2 - 2x - 3) - (x - 8)}{x^2 - 2x - 3} = \frac{2x^2 - 5x + 2}{(x - 3)(x + 1)}$$

$$= \frac{(2x - 1)(x - 2)}{(x - 3)(x + 1)} \; ,$$

and the denominator is

$$\frac{12x + 1}{x + 3} - \frac{6x - 17}{x - 3} = \frac{(12x + 1)(x - 3) - (6x - 17)(x + 3)}{(x + 3)(x - 3)}$$

$$= \frac{6x^2 - 36x + 48}{(x + 3)(x - 3)} = \frac{6(x - 4)(x - 2)}{(x + 3)(x - 3)} \; .$$

Therefore the complex fraction is

$$\frac{(2x - 1)(x - 2)}{(x - 3)(x + 1)} \cdot \frac{(x + 3)(x - 3)}{6(x - 4)(x - 2)} \; ,$$

which simplifies to

$$\frac{(2x - 1)(x + 3)}{6(x + 1)(x - 4)} \; .$$

EQUATIONS

Solve each of the following equations for the letter that follows it, and simplify the resulting expression.

1. $(x + 2)(y + \frac{5}{2}) = 80$; y

2. $\frac{2}{3}(x + 1) - \frac{1}{2}(y - \frac{2}{5}) = \frac{7}{8}$; y

3. $y = \frac{x + 3}{x - 4}$; x

4. $2xy + x^2y' - 3y^2y' = 0$; y'

5. $2 + \frac{2 - x}{x + 2} = \frac{3x + 7}{x + 5} - 2$; x

Simplify each of the following expressions.

6. $\dfrac{\frac{x + 1}{x} - 2}{x - 1}$

7. $\dfrac{2y - 4\left(\frac{2x}{3y^2}\right)}{y^3}$

8. $F(F(x))$, where $F(x) = \dfrac{x}{x + 1}$

9. $\dfrac{\frac{3}{x - 4} - \frac{16}{x - 3}}{\frac{2}{x - 3} - \frac{15}{x + 5}}$

10. Solve for x: $\dfrac{(x + 2)^3 - x(x + 2)(x + 4)}{(x + 2)^4} = 0$

ANSWERS

1. $y = \dfrac{5(30 - x)}{2(x + 2)}$

2. $y = \dfrac{4}{3}x - \dfrac{1}{60}$

3. $x = \dfrac{4y + 3}{y - 1}$

4. $y' = \dfrac{-2xy}{x^2 - 3y^2} = \dfrac{2xy}{3y^2 - x^2}$

5. $x = -3$

6. $-1/x$

7. $\dfrac{6y^3 - 8x}{3y^5} = 2\left(\dfrac{3y^2 - 4x}{3y^5}\right)$

8. $\dfrac{x}{2x + 1}$

9. $\dfrac{x + 5}{x - 4}$

10. The equation has no solution. ($x = -2$ is not a solution because the equation is undefined when $x = -2$.)

PART III: Algebra for Applications

When you apply the derivative you will need to solve equations and inequalities and systems of equations. These topics are reviewed in the first four sections. For applications of both the derivative and the integral, it is useful to know something about the conic sections. Thus, in the last section we review the standard algebraic forms for the circle, ellipse, parabola, and hyperbola. This is preceded by a section on writing algebraic expressions in prescribed forms, a skill useful in the study of integration techniques.

Many beginning calculus students regard the applications portion of the calculus course as the most difficult (and most important). The difficulties are due to the fact that applications are given in verbal form, and these must first be translated into equivalent algebraic problems. To help prepare you for this, we have included a section on translating verbal statements into algebraic expressions (Section 22) and a section showing how functions arise in applications (Section 23). Many applications require a knowledge of the elementary mensuration formulas, which are reviewed in Section 21. Section 24 reviews the conversion formulas needed to relate the English system of measurement with the metric system.

DIAGNOSTIC TEST: Over Part III

(Answers on page 182)

1. Solve for x: $\dfrac{x^2 + 4x - 1}{3 - 2x} = x$.

2. Solve for x: $x(x^2 - 13) = -12$.

3. Find all values of x for which $(x + 1)(x - 3)^2(2x - 1) \geq 0$.

4. Find numbers A and B such that for all x, $x + 2 = A(x - 3) + B(x + 1)$.

5. The surface of a sphere is 16π. Find the volume.

6. Because of traffic conditions, you can average 50 miles per hour in going from A to B but only 30 miles per hour on the return trip from B to A (same distance). (a) Would you guess your average velocity for the round trip to be more or less than 40 miles per hour? (b) Let d denote the distance between A and B, and let T_1 and T_2 denote the time from A to B and B to A respectively. What is the relationship between d and T_1? (d and T_2?) (c) Use the results in part (b) to express $T_1 + T_2$ in terms of d, and compute the average velocity for the round trip $(2d/(T_1 + T_2))$.

7. An isosceles triangle 10 feet across the top and 20 feet deep is partially submerged in water as shown at the right. Express the area of the submerged portion as a function of the depth h.

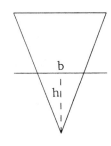

8. Express $y + 2x^2 - 5x + 1 = 0$ in the form $y = A(x - h)^2 + k$ and sketch its graph.

70

17. Quadratic Equations

An equation that can be put into the standard form

$$ax^2 + bx + c = 0$$

is called a *quadratic equation*. In some cases these equations can be solved by factoring as in the following example.

Example 1. Put the equation $x(x - 4) + 6 = (3x - 8)x$ into the standard form for a quadratic equation and find the solutions by factoring.

Solution. Expanding, we get

$$x^2 - 4x + 6 = 3x^2 - 8x,$$

$$-2x^2 + 4x + 6 = 0,$$

which is equivalent to

$$x^2 - 2x - 3 = 0.$$

Factoring the left side gives

$$(x - 3)(x + 1) = 0.$$

Because a product is zero if and only if at least one of its factors is zero, it follows that either $x - 3 = 0$ or $x + 1 = 0$. The solutions are therefore $x = 3$ and $x = -1$.

The solution to the general quadratic equation $ax^2 + bx + c = 0$ for all possible parameters a, b, c   (a ≠ 0) is given by the *quadratic formula*

$$x = \frac{-b \pm \sqrt{b^2 - 4ac}}{2a}.$$

To be completely general this formula must allow for the possibility of complex numbers as solutions to certain equations.

71

Applying the quadratic formula to the equation $x^2 - 2x - 3 = 0$ (from Example 1), we have $a = 1$, $b = -2$, and $c = -3$, and we find the solutions to be

$$x = \frac{2 \pm \sqrt{4 - 4 \cdot 1 \cdot (-3)}}{2} = \frac{2 \pm \sqrt{16}}{2} = \frac{2 \pm 4}{2}$$

$$= \frac{6}{2}, \quad \frac{-2}{2} = 3, -1.$$

Note the fundamental difference between the method of solving linear equations (Sections 5, 6, and 16) and the method of solving quadratic equations. In the case of linear equations, we solve by collecting the constants on one side and the terms containing the variable on the other. This procedure doesn't work when applied to quadratic equations; for example, writing $6x^2 + 7x - 3 = 0$ in the form $x(6x + 7) = 3$ leads nowhere. Rather, the first step in solving a quadratic equation is to bring it into standard form, to solve by factoring if the factors are easily found, or to substitute the proper values of a, b, c into the quadratic formula.

Many equations can be transformed into quadratic form by algebraic manipulation; the next two examples are typical.

Example 2. Find all values of x which satisfy $\dfrac{x + 2}{x + 1} + \dfrac{x - 1}{x - 3} = \dfrac{19}{6}$.

Solution. $6[(x + 2)(x - 3) + (x - 1)(x + 1)] = 19(x + 1)(x - 3)$

$$6[(x^2 - x - 6) + (x^2 - 1)] = 19(x^2 - 2x - 3)$$

$$6(2x^2 - x - 7) = 19(x^2 - 2x - 3)$$

$$12x^2 - 6x - 42 = 19x^2 - 38x - 57$$

$$7x^2 - 32x - 15 = 0$$

Substituting $a = 7$, $b = -32$, $c = -15$ into the quadratic formula we have

$$x = \frac{32 \pm \sqrt{(32)^2 - 4 \cdot 7(-15)}}{14}$$

72

$$= \frac{32 \pm \sqrt{1444}}{14} = \frac{32 \pm 38}{14} .$$

Thus, there are two solutions, $x = 70/14 = 5$ and $x = -6/14 = -3/7$. (Neither of them is an *extraneous* root: see the discussion following Example 2, Section 16.) The solutions could have been found without recourse to the quadratic formula if we had noticed that $7x^2 - 32x - 15$ factors into $(7x + 3)(x - 5)$.

<u>Example 3.</u> Find all values of x for which $\sqrt{x + 1} + \sqrt{x + 8} = \sqrt{5x + 9}$.

<u>Solution.</u> We first eliminate square roots. This is done by squaring twice, but with a necessary intermediate step. Squaring each side once gives

$$(x + 1) + 2\sqrt{(x + 1)(x + 8)} + (x + 8) = 5x + 9,$$

$$(2x + 9) + 2\sqrt{(x + 1)(x + 8)} = 5x + 9.$$

Before squaring the second time, we isolate the remaining square root (or else squaring the second time will not eliminate the square root). Thus, we re-arrange the last equation and put it into the form

$$2\sqrt{(x + 1)(x + 8)} = 3x.$$

Now squaring each side a second time yields

$$4(x + 1)(x + 8) = 9x^2,$$

$$4(x^2 + 9x + 8) = 9x^2.$$

This equation written in standard quadratic form is equivalent to

$$5x^2 - 36x - 32 = 0.$$

The quadratic formula yields

$$x = \frac{36 \pm \sqrt{(36)^2 - 4 \cdot 5(-32)}}{10}$$

$$= \frac{36 \pm \sqrt{1936}}{10} = \frac{36 \pm 44}{10} .$$

Thus, the solutions to the *quadratic equation* are x = 8 and x = -4/5. These correspond to those we would have found had we factored $5x^2 - 36x - 32$ as (5x + 4)(x - 8). You can check that x = 8 is a solution to the original problem; however, x = -4/5 is not. This example shows that the original equation and the squared equation may not be equivalent. Squaring each side of an equation may introduce *extraneous* roots. For example, x = -1 has only one root, but $x^2 = 1$ has two roots. Therefore all roots obtained in this way must be checked.

EXERCISES

Find all values of x which satisfy the following equations.

1. $3x^2 + 2x - 1 = 0$

2. $x + x^2 = 2$

3. $x^2 - (x - 2)2x = 4$

4. $3x^2 - 4x - 1 = 0$

5. $x = \dfrac{4}{4 - x}$

6. $\dfrac{3}{x - 2} + \dfrac{x}{x + 2} = \dfrac{8}{x^2 - 4}$

7. $\sqrt{8 - x^2} - \dfrac{x^2}{\sqrt{8 - x^2}} = 0$

8. $\dfrac{-5/x^2}{1 + (5/x)^2} - \dfrac{-3/x^2}{1 + (3/x)^2} = 0$

9. $\sqrt{2x - 2} + \sqrt{x - 5} = 2\sqrt{x}$

10. $x^2 + 4cx + c^2 = 2$

ANSWERS

1. x = -1, x = 1/3   2. x = 1, x = -2   3. x = 2   4. $x = \dfrac{2 \pm \sqrt{7}}{3}$

5. x = 2   6. x = 1 (x = -2 is not a solution)   7. x = ±2

8. $x = \pm\sqrt{15}$ 9. x = 9   10. $x = -2c \pm \sqrt{3c^2 + 2}$

18. Polynomial Equations of Higher Degree

In previous sections, we have shown how to solve linear and quadratic equations. It is much harder to solve polynomial equations of higher degree. The search for roots often proceeds by trial and error.

Example 1. Find all solutions of $x^3 - 2x^2 - 5x + 6 = 0$.

Solution. By direct substitution, we discover that 1 is a root $[(1)^3 - 2(1)^2 - 5(1) + 6 = 0)]$. We next divide $x^3 - 2x^2 - 5x + 6$ by $x - 1$:

$$
\begin{array}{r}
x^2 - x - 6 \\
x - 1 \overline{\smash{\big)}\ x^3 - 2x^2 - 5x + 6} \\
\underline{x^3 - x^2 } \\
-x^2 - 5x \\
\underline{-x^2 + x } \\
-6x + 6 \\
\underline{-6x + 6}
\end{array}
$$

This division is exact, and therefore $x^3 - 2x^2 - 5x - 6 = (x - 1)(x^2 - x - 6)$. In this case, the quadratic term factors nicely, and we find that the original polynomial equation is equivalent to

$$(x - 1)(x - 3)(x + 2) = 0.$$

From this, we find that the solutions must be $x = 1$, $x = 3$, and $x = -2$.

Example 2. Find all values of x which satisfy

$$x^5 - 2x^4 - 5x^3 + 10x^2 + 6x - 12 = 0.$$

Solution. The only possible integer roots for this equation are ± 1, ± 2, ± 3, ± 4, ± 6, ± 12. (This follows from the more general fact that if an integer k is a root of the polynomial equation $x^n + a_{n-1}x^{n-1} + \ldots + a_1 x + a_0 = 0$, with integer coefficients, then k must divide the constant term a_0.) Direct substitution of these possible roots shows that 2 is a root. Dividing the polynomial

by x - 2 yields

$$
\begin{array}{r}
x^4 \qquad - 5x^2 \qquad + 6 \\
x - 2 \overline{)\; x^5 - 2x^4 - 5x^3 + 10x^2 + 6x - 12} \\
\underline{x^5 - 2x^4} \qquad\qquad\qquad\qquad \\
- 5x^3 + 10x^2 \qquad\qquad \\
\underline{- 5x^3 + 10x^2} \qquad\qquad \\
6x - 12 \\
\underline{6x - 12}
\end{array}
$$

Therefore, the original equation is equivalent to

$$(x - 2)(x^4 - 5x^2 + 6) = 0,$$

$$(x - 2)(x^2 - 3)(x^2 - 2) = 0.$$

In this form we are able to read off all the solutions: $x = 2$, $x = \pm\sqrt{3}$,

$x = \pm\sqrt{2}$.

We would have arrived at this same point had we noticed that the terms in

the original problem could be grouped into the form

$$(x^5 - 2x^4) + (-5x^3 + 10x^2) + (6x - 12) = 0,$$

$$x^4(x - 2) - 5x^2(x - 2) + 6(x - 2) = 0,$$

$$(x^4 - 5x^2 + 6)(x - 2) = 0,$$

$$(x^2 - 3)(x^2 - 2)(x - 2) = 0.$$

Example 3. Find all roots of $3x^4 + 13x^3 + 10x^2 + 2x = 0$.

Solution. Obviously, 0 is a root. Dividing this out, the other roots must

satisfy

$$3x^3 + 13x^2 + 10x + 2 = 0.$$

The only possible rational roots are ± 1, ± 2, $\pm 1/3$, $\pm 2/3$. (This follows from

the Rational Root Theorem, which states that if r is a rational number which is

a root of the polynomial equation $a_n x^n + \ldots + a_1 x + a_0 = 0$, with integer coefficients, then r can be written in the form p/q, where p and q are integers, p divides the constant term a_0, and q divides the leading coefficient a_n.) If each of these potential rational roots is checked, we find that only $x = -1/3$ satisfies the equation. Dividing by $x + 1/3$ gives

$$
\begin{array}{r}
3x^2 + 12x + 6 \\
x + \dfrac{1}{3} \overline{\smash{\big)}\ 3x^3 + 13x^2 + 10x + 2} \\
\underline{3x^3 + x^2} \\
12x^2 + 10x \\
\underline{12x^2 + 4x} \\
6x + 2 \\
\underline{6x + 2}
\end{array}
$$

so that the original equation factors into the equivalent form

$$ x(x + \frac{1}{3})(3x^2 + 12x + 6) = 0. $$

The final two roots are found by solving the quadratic equation

$$ x^2 + 4x + 2 = 0. $$

The preceding examples illustrate one procedure for solving polynomial equations of higher degree: namely, factor the polynomial either by grouping the terms in a clever way, or by finding a root by direct substitution. If r is a root of $P(x) = 0$, $P(x)$ a polynomial of degree n, $n > 2$, then $P(x) = (x - r)Q(x)$ for some polynomial $Q(x)$ of degree $n - 1$. The polynomial $Q(x)$ can be found by dividing $P(x)$ by $x - r$. Therefore, if we know one root, say r, of $P(x) = 0$, then the other roots can be found by solving $Q(x) = 0$. We have made the problem easier in the sense that we have reduced the degree of the polynomial by one. This entire process can be repeated until finally the problem is reduced to a quadratic equation.

Find all solutions to the following equations.

1. $x^3 - 4x + 3 = 0$

4. $(x^3 - 2x^2) + (-x + 2) = 0$

2. $x^3 - 4x^2 + x + 6 = 0$

5. $x^5 - 5x^3 + 4x = 0$

3. $x^3 + x^2 - x - 1 = 0$

6. $x^3 - 9x + x^2 - 9 = 0$

7. $2(x + 3)(x - 5)^2 + 2(x + 3)^2(x - 5) = 0$ (Begin by factoring.)

8. $(x - 3)^2(5x - 3) + 2(x + 1)(x - 3)(5x - 3) + (x + 1)(x - 3)^2(5) = 0$

9. $\dfrac{(x^2 + 1)^2(-2x) - (1 - x^2)2(x^2 + 1)2x}{(x^2 + 1)^4} = 0$

10. By long division show that $\dfrac{x^4 + 1}{(x + 1)(x - 2)^2} = x + 3 + \dfrac{9x^2 - 4x - 11}{(x + 1)(x - 2)^2}$.

ANSWERS

1. $x = 1, \dfrac{-1 \pm \sqrt{13}}{2}$ 2. $x = -1, 2, 3$ 3. $x = 1, -1$

4. $x = -1, 1, 2$ 5. $x = \pm 2, \pm 1, 0$ 6. $x = -3, -1, 3$

7. $x = -3, 1, 5$ 8. $x = 3, \dfrac{3 \pm 2\sqrt{6}}{5}$ 9. $x = 0, \pm\sqrt{3}$

---

19. Inequalities

In this section we will consider certain nonlinear inequalities and extend the ideas introduced in Section 9.

Example 1. Find all real numbers x which satisfy the equation $x(x - 1) \le 6$.

Solution. We proceed exactly as though this were a quadratic equation (rather than a quadratic inequality) by bringing all terms to the left side:

$$x^2 - x - 6 \leq 0,$$

and express it in factored form as

$$(x - 3)(x + 2) \leq 0.$$

This suggests dividing the line into three parts as shown below.

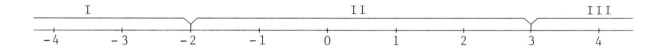

When x is in region I, both $(x - 3)$ and $(x + 2)$ are negative, and there-fore the product $(x - 3)(x + 2)$ is positive.

When x is in region II, $(x + 2)$ is positive whereas $(x - 3)$ is negative, so that $(x - 3)(x + 2)$ is negative.

When x is in region III, $(x - 3)(x + 2)$ is positive because each factor is positive.

Therefore, $(x - 3)(x + 2) \leq 0$ if and only if $-2 \leq x \leq 3$.

Example 2. Find all real numbers x which satisfy $\dfrac{3x}{x - 1} < \dfrac{x}{x + 2} + 2$.

Solution. If this were an equation rather than an inequality, the first step would be to multiply by $(x - 1)(x + 2)$ to eliminate denominators. However, any attempt to clear denominators will necessarily mean dividing the problem into two cases, depending upon whether $(x - 1)(x + 2)$ is positive or negative.

Another approach is to proceed much as we did in the last example. Bring everything to the left side,

$$\frac{3x}{x - 1} - \frac{x}{x + 2} - 2 < 0,$$

and simplify

$$\frac{3x(x + 2) - x(x - 1) - 2(x - 1)(x + 2)}{(x - 1)(x + 2)} < 0$$

$$\frac{3x^2 + 6x - x^2 + x - 2x^2 - 2x + 4}{(x - 1)(x + 2)} < 0$$

$$\frac{5x + 4}{(x - 1)(x + 2)} < 0.$$

Now, as in Example 1, divide the line into four regions by the points $x = -2$, $x = -4/5$, and $x = 1$. The table below gives the sign of each of the relevant factors in each region.

	$(x + 2)$	$(5x + 4)$	$(x - 1)$	$\dfrac{5x + 4}{(x - 1)(x + 2)}$
I. $x < -2$	−	−	−	−
II. $-2 < x < -\dfrac{4}{5}$	+	−	−	+
III. $-\dfrac{4}{5} < x < 1$	+	+	−	−
IV. $x > 1$	+	+	+	+

From this table we find that the solution is all real numbers that are either less than −2 or strictly between − 4/5 and 1: $x < -2$ or $-4/5 < x < 1$.

EXERCISES

Find all real numbers x which satisfy each of the following inequalities.

1. $x^2 - x - 12 < 0$

2. $x(x + 4) < -3$

3. $(x - 1)^2(x - 2)^4 > 0$

4. $x(x - 1)^2(x - 2)^3 > 0$

5. $\dfrac{(x - 4)^3(x + 2)}{(x - 1)^2} \geq 0$

6. $\dfrac{1}{x - 1} < 1$

7. $\dfrac{x - 3}{x + 1} < 2$

8. $\dfrac{2x + 1}{x} > \dfrac{2x - 1}{x - 1}$

9. $(2x + 1)(x^3 + 3) < x(2x + 1)(3x + 1)$

10. $\dfrac{1}{3} x^{-2/3}(x - 7)^2 + 2x^{1/3}(x - 7) > 0$

80

1. $-3 < x < 4$ 2. $-3 < x < -1$ 3. All reals except 1 and 2

4. $x < 0$ or $x > 2$ 5. $x \leq -2$ or $x \geq 4$ 6. $x < 1$ or $x > 2$

7. $x < -5$ or $x > -1$ 8. $0 < x < 1$ 9. The equation is equivalent to

$(x - 1)(x + 1)(2x + 1)(x - 3) < 0$; $-1 < x < -1/2$ or $1 < x < 3$.

10. $x < 0$ or $0 < x < 1$, or $x > 7$.

---

20.   Systems of Equations

The solution to certain problems requires finding numbers which satisfy two equations simultaneously. Geometrically, these solutions (if they exist) correspond to points on the intersections of the two respective graphs.

<u>Example 1</u>. Find all values of x and y which simultaneously satisfy both the equations

$$y = 3 - x^2$$

$$y = 3 - 2x.$$

<u>Solution</u>. Substituting y from the second equation into the first, we have

$$3 - 2x = 3 - x^2, \quad \text{or equivalently,} \quad x^2 - 2x = 0.$$

It follows that $x = 0$ or $x = 2$. Substituting these values for x back into either of the original equations, we find the corresponding y-coordinates: $y = 3$ when $x = 0$, and $y = -1$ when $x = 2$. That is, there are two solutions: $(0,3)$ and $(2,-1)$. These can be seen geometrically by carefully graphing the two equations.

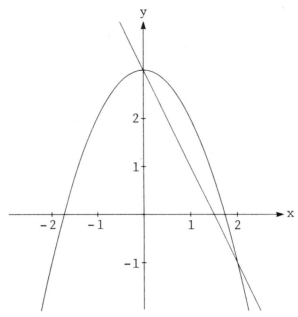

Example 2. Solve the system of equations $9x + 2y = 37$

$$5x + 6y = 45.$$

Solution. The problem asks us to find the points which satisfy both equations simultaneously. The equations correspond to nonparallel straight lines, and therefore there will be exactly one solution (at the intersection point). One approach for finding this point is to solve for one of the variables in one of the equations, and to substitute the resulting expression into the other equation. Thus, solving for y in the first equation,

$$y = \frac{37 - 9x}{2},$$

and substituting this into the second equation,

$$5x + 6\left(\frac{37 - 9x}{2}\right) = 45,$$

$$5x + 3(37 - 9x) = 45,$$

$$5x + 111 - 27x = 45,$$

$$-22x = -66$$

$$x = 3.$$

It follows that the y-coordinate at the point of intersection is 3 and that

$$y = \frac{37 - 9x}{2} = \frac{37 - 27}{2} = 5.$$

The point of intersection therefore is (3,5).

Another method for finding the intersection point is to multiply one or both equations by appropriate constants so that when the two are added, one of the variables is eliminated. Thus, for example, multiply the first equation by -3 to get

$$-27x - 6y = -111,$$

$$5x + 6y = 45.$$

Adding these equations (the y-terms add to zero), we get

$$-22x = -66,$$

$$x = 3.$$

To find the y coordinate of the intersection point, multiply the first of the original equations by -5 and the second by 9, to obtain

$$-45x - 10y = -185,$$

$$45x + 54y = 405.$$

Adding these equations (the x-terms add to zero), we get

$$44y = 220,$$

$$y = 5.$$

Example 3. Find numbers A, B, and C such that

$$\frac{x^2 + 3x + 1}{(x + 4)(x^2 + 1)} = \frac{A}{x + 4} + \frac{Bx + C}{x^2 + 1}.$$

Solution. Adding the terms on the right side we get

$$\frac{x^2 + 3x + 1}{(x + 4)(x^2 + 1)} = \frac{A(x^2 + 1) + (Bx + C)(x + 4)}{(x + 4)(x^2 + 1)}.$$

Multiplying by $(x + 4)(x^2 + 1)$ gives

$$x^2 + 3x + 1 = A(x^2 + 1) + (Bx + C)(x + 4).$$

This equation must hold for all values of x (except x = -4 when the original equation is undefined); in particular it holds for x = -1, x = 0, and x = 1. Substituting these values into the last equation, we get the three equations

$$-1 = 2A - 3B + 3C,$$

$$1 = A + 4C,$$

$$5 = 2A + 5B + 5C.$$

From the second equation, A = 1 - 4C, and substituting this into the first and third equations, we get

$$-1 = 2(1 - 4C) - 3B + 3C,$$

$$5 = 2(1 - 4C) + 5B + 5C,$$

or, equivalently,

$$-3 = -3B - 5C,$$

$$3 = 5B - 3C.$$

Proceeding as in Example 1 we can show that the solution to these equations is B = 12/17, C = 3/17. For these values of B and C, we get A = 1 - 4C = 1 - 4(3/17) = 5/17.

EXERCISES

Solve the following systems of equations.

1. $-x + y = -1$

    $x + y = 3$

2. $3x + y = 10$

    $4x + 5y = -16$

3. $3x + 5y = 11$

    $8x - 12y = -15$

4. $y = 3 - x^2$

    $y = -2x$

5. $y = 5 - x^2$

    $y = x^2 - 3$

6. $x + y = 3$

    $xy = 2$

7. $y = 2x^2$

    $y = x^2 + 3x + 4$

8. $\dfrac{y}{x - a} = -\dfrac{b}{c}$

    $\dfrac{y}{x} = \dfrac{a - b}{c}$

9. Find the numbers A and B such that $\dfrac{2x + 8}{(x - 2)(x + 1)} = \dfrac{A}{x - 2} + \dfrac{B}{x + 1}$.

10. Find numbers A, B, and C such that $\dfrac{5x^2 + 4x + 3}{(x + 1)(x^2 + 1)} = \dfrac{A}{x + 1} + \dfrac{Bx + C}{x^2 + 1}$.

ANSWERS

1. $x = 2$, $y = 1$ 2. $x = 6$, $y = -8$ 3. $x = 57/76$, $y = 133/76$

4. $x = 3$, $y = -6$; $x = -1$, $y = 2$ 5. $x = 2$, $y = 1$; $x = -2$, $y = 1$

6. $x = 1$, $y = 2$; $x = 2$, $y = 1$ 7. $x = -1$, $y = 2$; $x = 4$, $y = 32$

8. $x = b$, $y = \dfrac{b}{c}(a - b)$ 9. $A = 4$, $B = -2$ 10. $A = 2$, $B = 3$, $C = 1$

21. Geometric Prerequisites

Here are several area and volume formulas from geometry that you will need to know for the calculus course.

<u>Rectangle:</u>   Base b, Height h

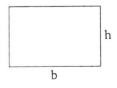

    Perimeter: $P = 2(b + h)$

    Area: $A = bh$

Parallelogram: Base b, Height h

    Area: $A = bh$

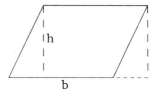

Triangle: Base b, Height h

    Area: $A = \frac{1}{2} bh$

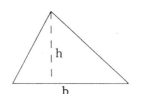

Trapezoid: Bases b_1 and b_2, Height h

    Area: $A = \frac{1}{2} (b_1 + b_2)h$

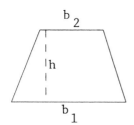

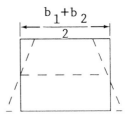

Circle: Radius r

    Area: $A = \pi r^2$

    Circumference: $C = 2\pi r$

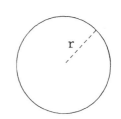

Rectangular Box: Length ℓ, Width w, Height h

    Volume: $V = \ell wh$

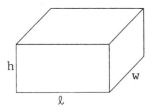

Right Circular Cylinder: Radius r, Height h

    Volume: $V = \pi r^2 h$

    Surface Area of
      Lateral Side: $A_s = 2\pi rh$

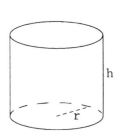

Right Circular Cone: Radius r, Height h

Volume: $V = \frac{1}{3} \pi r^2 h$

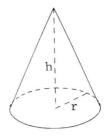
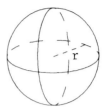

Sphere: Radius r

Volume: $V = \frac{4}{3} \pi r^3$

Surface Area: $S = 4\pi r^2$

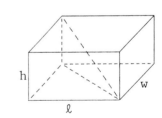

Two important geometrical facts that are very helpful in the analysis of applied calculus problems (particularly max-min problems and related rates problems) are:

      (i) Pythagorean Theorem,

      (ii) Corresponding sides of similar triangles are proportional.

Example 1. Write a formula for the length L of the diagonal of a rectangular box.

Solution. Let w, ℓ, and h denote the width, length, and height of the box. By the Pythagorean Theorem, the length of the base diagonal is $\sqrt{w^2 + \ell^2}$. Applying the Pythagorean Theorem a second time, we find that $L = \sqrt{w^2 + \ell^2 + h^2}$.

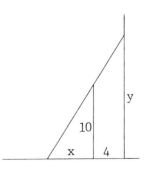

Example 2. A ladder leans over a 10 foot fence against a house 4 feet away from the fence (see figure). Find a relationship between y, the height of the top of the ladder above the

87

ground, and x, the distance of the foot of the ladder from the fence.

Solution. Using the fact that corresponding parts of similar triangles are proportional, we have

$$\frac{4 + x}{y} = \frac{x}{10}.$$

Example 3. Imagine that the earth is a smooth sphere and that a string is wrapped around it at the equator. Now suppose that the string is lengthened by six feet and the new length is evenly pushed out to form a larger circle just over the equator. Estimate the distance between the string and the surface of the earth. Is it more or less than one inch?

Solution. Let R denote the radius of the earth, and R' denote the radius of the larger circle formed after the string has been lengthened by six feet. On the one hand, the circumference of the string is $2\pi R'$; on the other hand, its circumference is $2\pi R + 6$. Thus

$$2\pi R' = 2\pi R + 6,$$

and from this we find that

$$R' - R = \frac{6}{2\pi}.$$

Note: This equation means that the distance between the string and the earth is $6/2\pi$, or nearly one foot! The distance is independent of the radius R; the same result would have been obtained had we used a globe or even a golf ball.

EXERCISES

1. Write a formula that gives the surface area of a rectangular box.

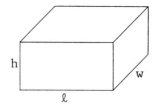

2. A Norman window consists of a rectangle surmounted by a semicircle. Write a formula for the perimeter of the window.

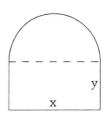

3. A silo consists of a cylinder surmounted by a hemisphere. Write a formula for the surface area (side and top).

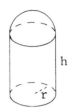

4. A cylindrical tin kettle, open at the top, has a copper bottom. If copper is five times as expensive as tin (per square area), write a formula for the cost of the material.

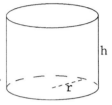

5. A lot has the form of a right triangle, with perpendicular sides 90 and 120 feet long. A rectangular building of length x and width y is to be erected on the corner facing the perpendicular sides. Use similar triangles to find a relationship between x and y.

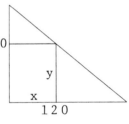

6. A rectangle with base x and height y is inscribed in an isosceles triangle with base 10 and height 20. Find a relationship between x and y.

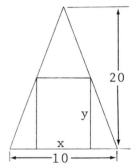

7. An equilateral triangle has side length equal to 10. Find its height.

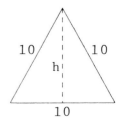

8. A rectangle of width x and height y is inscribed in a

circle of radius 5. Find a relationship between x and

y.

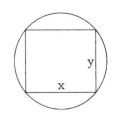

9. A cone of base radius r and height h is inscribed in a

sphere of radius 5. Find a relationship between r and

h.

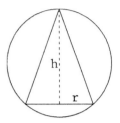

10. A one mile length of railroad track is

laid on a flat stretch of ground in the

middle of winter; it is fastened securely

at both ends. Suppose that during the summer the track expands two feet

and buckles. Estimate the height above the ground at the highest point.

Is it more or less than ten feet? (Hint: The problem does not indicate

what shape the buckled track will assume. If we assume the track is part

of a circle, the mathematics is pretty hairy! Since we want only an esti-

mate, suppose the shape is that of an isosceles triangle (see figure).)

ANSWERS

1. $A_s = 2[wh + w\ell + h\ell]$ 2. $P = x + 2y + \pi x/2$ 3. $A_s = 2\pi rh + 2\pi r^2$

4. $C = k[5\pi r^2 + 2\pi rh]$, k = cost per unit square area of tin

5. $\dfrac{120 - x}{y} = \dfrac{120}{90}$, or equivalently, $\dfrac{x}{90 - y} = \dfrac{120}{90}$, or equivalently,

$\dfrac{x}{90 - y} = \dfrac{120 - x}{y}$.   (In this problem, and in #6, other equivalent answers are

possible.)   6. $\dfrac{x}{20 - y} = \dfrac{10}{20}$, or equivalently, $\dfrac{5 - x/2}{y} = \dfrac{5}{20}$

7. $h = \sqrt{100 - 25} = 5\sqrt{3}$ 8. $x^2 + y^2 = 100$

9. $r^2 + (h - 5)^2 = 25$ 10. Let h denote the height of the isosceles

90

triangle. Then, $h = \sqrt{(2641)^2 - (2640)^2} = \sqrt{(2641 - 2640)(2641 + 2640)} =$

$\sqrt{5281} \cong 72.7$. From this computation, it is reasonable to assume that the

highest point of the track, even regardless of the shape, is more than ten feet.

22. Translation into Algebraic Notation

Applied calculus problems are usually given in verbal form (word problems).
It is therefore important to be able to transform verbal descriptions of relationships into equivalent algebraic equations.

Example 1.   The reciprocal of the focal length f of a thin lens is equal to
the sum of the reciprocals of the object distance u and the image distance v.
Express this relationship algebraically.   Find the image distance v when the
focal length is 20 centimeters and the object distance is 60 centimeters.

Solution.   The equation is

$$\frac{1}{f} = \frac{1}{u} + \frac{1}{v} \, .$$

(The "is" or "is equal to" in a verbal statement translates into an equal sign;
this is a useful clue in picking out the left and right sides of the equation.)
When f = 20 and u = 60, we have

$$\frac{1}{20} = \frac{1}{60} + \frac{1}{v} \, ,$$

$$\frac{1}{20} = \frac{v + 60}{60v} \, ,$$

$$3v = v + 60,$$

$$2v = 60,$$

$$v = 30.$$

A common phrase found in verbal statements relating variables is "is proportional to." Such a phrase translates into "is equal to a constant k times." More generally, the statements

y is proportional to x,

y varies as x,

y varies directly as x,

are all equivalent and translate into the equation

$$y = kx.$$

In such a relationship, a change in x causes a proportional change in y; for example, when x is doubled, so also is y. This fact does not depend on the particular value of the constant k, which is called the *constant of proportionality*.

Example 2. The kinetic energy E of a moving body is proportional to the square of its velocity v. Express this relationship algebraically and compare the kinetic energy of a man jogging at 5 miles per hour with his kinetic energy when sprinting at 20 miles per hour.

Solution. In symbols, $E = kv^2$. The kinetic energies while jogging, E_j, and while sprinting, E_s, are given by

$$E_j = k(5)^2 = 25 k, \quad \text{and} \quad E_s = k(20)^2 = 400 k.$$

It follows that $E_s/E_j = 16$, which is to say that his kinetic energy when sprinting is sixteen times greater than when jogging.

A related phrase that appears often in science courses is "inversely proportional." More precisely, the statements

y is inversely proportional to x,

y varies inversely as x,

are equivalent, and translate into the equation

$$y = \frac{k}{x} \,.$$

Notice that, in this equation, when x is doubled, y is halved.

Example 3.   Newton's law of gravitational attraction states that the force F with which two particles of mass m_1 and m_2 attract each other is proportional to the product of their masses and inversely proportional to the square of the distance r between the particles.   Write this in the form of an equation.

Solution.   Translating into symbols, we have $F = k \dfrac{m_1 m_2}{r^2} \,.$

An understanding of the relationship between velocity, distance, and time is crucial for many applied calculus problems.   Suppose that an object traverses a distance d when moving with constant velocity v during a time interval t.   The relationship between d, v, and t is given by the formula

$$v = \frac{d}{t} \,.$$

This formula is easy to remember if you simply think about the units of measurement of each quantity in the formula.   *Velocity* is measured in miles per hour, feet per second, kilometers per minute, etc., each of which is a measure of *distance* divided by a measure of *time*.

EXERCISES

1.   The temperature C in degrees Celsius (centigrade) is equal to 5/9 times the number found by subtracting 32 from the temperature F in degrees Fahrenheit. a) Express this symbolically.   b) What temperature F in degrees Fahrenheit corresponds to 20 degrees Celsius?

2.   According to Bernoulli's law, if the pressure in a water pipe is 100 pounds

per square inch when the water is not flowing, then when the water is flowing the pressure p (pounds per square inch) plus twice the square of the velocity of the water v (feet per second) is equal to 100. Express this relationship algebraically.

3. An investment earns the total simple interest I equal to P, the initial amount invested, times the product of the interest rate r (in percent per year) and the time t (in years) since the initial investment. a) Express this algebraically. b) In two years, how much simple interest does an investment of $1000 earn at 7 percent per year (r = .07, t = 2)? c) What initial investment will yield $10 interest every three months if the interest rate is 5 percent per year (r = .05, t = 1/4)?

4. If the temperature of a gas (such as air) is held constant, its pressure P is inversely proportional to its volume V (Boyle's Law). a) Express this as a formula. b) Find the constant of proportionality, if it is known that the pressure is 150 pounds per square inch when the volume is 600 cubic inches. c) Suppose the volume is increased to 900 cubic inches. What will be the pressure if there is no change in temperature?

5. The distance d in miles that a person can see to the horizon from a point h feet above the surface of the earth is approximately proportional to the square root of the height h. a) Express this relationship algebraically. b) Find the constant of proportionality if it is known that the horizon is 30 miles away at a height of 400 feet. c) Approximately how far is the horizon from a point 900 feet high?

6. The illumination I in foot-candles upon a wall is proportional to the intensity i in candlepower of the source of light and inversely proportional as the square of the distance d of the wall from the light. a) Express

this relationship algebraically. b) If the illumination is 6 foot-candles at a distance 10 feet from a light of 600 candlepower, what is the illumination at a distance of 15 feet from a light of 3000 candlepower?

7. The time of exposure t necessary to photograph an object is proportional to the square of the distance d of the object from the light source and inversely proportional to the intensity of illumination I. a) Express this relationship algebraically. b) Suppose that the correct exposure is 1/50 seconds when the light is 5 feet from the object. What must be the distance of the object from the light source if the light intensity is doubled and the exposure time is increased to 1/10 seconds? (Hint: Solve for k/I.)

8. The general gas law states that the density ρ of equal volumes of different gases is proportional to the product of the absolute pressure P and the molecular mass M, and inversely proportional to the absolute temperature T. Express this as a formula.

9. Two cars start at the same place at the same time; one travels north at 40 mph, the others travels east at 30 mph. a) How far are they apart after 2 hours? b) How far are they apart after t hours?

10. An airplane cruises at x miles per hour in still air, but there is a wind of y miles per hour. With the wind, the airplane flies 36 miles (at x + y miles per hour) in the same time it takes to fly 30 miles against the wind (at x - y miles per hour). Express this relationship algebraically.

ANSWERS

1. a) $C = \frac{5}{9} (F - 32)$,   b) $F = 68°$ 2. $p + 2v^2 = 100$ 3. a) $I = Prt$,
b) $I = 140$,   c) $800   4. a) $P = k/V$,   b) $k = 90,000$ in-lb ,   c) $P = 100$ lb/in^2

95

5. a) $d \doteq k\sqrt{h}$,   b) $k = 3/2$,   c) $d \doteq 45$ ft   6. a) $I = ki/d^2$,

b) $I = 40/3$ ft-c   7. a) $t = kd^2/I$,   b) $d^2 = 250$, $d = 5\sqrt{10}$ ft   8. $\rho = k\dfrac{PM}{T}$

9. a) 100 mi,   b) 50t mi   10. Let T denote the time it takes to fly 36 miles with the wind, or 30 miles against the wind. Then $x + y = 36/T$ and $x - y = 30/T$. Solving each of these for T we get $\dfrac{36}{x + y} = \dfrac{30}{x - y}$.

23. Functions, Dependence

The definition of *function* was given in Section 15. In this section we will focus on what it means for one quantity to be a *function of* another quantity.

Whenever one quantity, for example y, depends upon another quantity, for example x, in such a way that y is uniquely determined once x is given, then we say that y is a *function of* x.

Consider the following illustrations:

(a) The area A of a circle is a function of the radius r, because once a positive value for r is given, the area A is uniquely determined.

(b) The area A of a circle is a function of its circumference C because once the circumference is given, the radius is uniquely determined and hence also the area.

(c) The circumference C of a circle is a function of the area A, for the area uniquely determines the radius, which in turn uniquely determines the circumference.

(d) If a ball is dropped from a tower, the distance it falls is a function of time.

(e) The distance in which a car can be stopped is a function of the velocity at which the car is traveling when the brakes are applied.

The feeling to be sensed is that of dependence, or cause and effect: knowing one variable uniquely determines another.

We express the fact that y is a function of x by writing $y = f(x)$, or $y = g(x)$, etc., where "f," "g," etc., denote the function. This equation is to be read as "y is a function of x." Sometimes the rule describing the function can be given explicitly as a formula, and sometimes it can't. Consider the five situations above.

(a) The area of a circle is a function of the radius, so we can write

$$A = f(r).$$

In this case, we can give an explicit formula for f, namely,

$$f(r) = \pi r^2.$$

The domain of f is the set of all positive numbers (corresponding to positive radii), and f assigns the area, πr^2, to r. It is customary to suppress the middle step containing the f and to write

$$A(r) = \pi r^2,$$

where we now think of A as the function, and A(r) as the value (area) of the function for the radius r.

(b) The area of a circle is a function of the circumference. The precise functional relationship is not immediately obvious. To find it, we begin by writing down what we know about the area and circumference (the two quantities of interest), introducing new variables as necessary. Thus, we know that

$$A = \pi r^2 \quad \text{and} \quad C = 2\pi r.$$

Our problem is to write A in terms of C; that is, we need to replace r in πr^2 by an expression containing C. Thus, from $C = 2\pi r$ we get $r = C/\pi 2$, and sub-

97

tituting this into $A = \pi r^2$ we get $A = \pi(\frac{C}{2\pi})^2$. This is the explicit formula we wanted, namely:

$$A(C) = (\frac{1}{4\pi})C^2.$$

(c) The circumference of a circle is a function of the area. To express this functional relationship explicitly, we begin with what we know about area and circumference:

$$A = \pi r^2 \quad \text{and} \quad C = 2\pi r.$$

Solving for r in the first equation,

$$r = \sqrt{\frac{A}{\pi}}$$

and substituting it into the second equation we get C as a function of A:

$$C(A) = 2\pi \sqrt{\frac{A}{\pi}} .$$

In (a), (b), and (c) we were able to express the function explicitly as a formula; but in (d) and (e) a formula for the value of the function cannot be written without knowing more about the laws of physics. The discovery of this relationship may require considerable experimentation.

Example 1. Express the area of a square as a function of its diagonal.

Solution. The problem is asking for a formula which gives the area of a square in terms of the diagonal. The variables of interest are area and diagonal length:

$$A = s^2 \quad \text{and} \quad d^2 = 2s^2 \ (= s^2 + s^2),$$

where s is the length of the side of the square. From the second equation, $s^2 = d^2/2$; therefore, the formula we want is

$$A(d) = \frac{d^2}{2} .$$

Example 2. A six-foot man walks past a 24 foot lamppost. Express the length of his shadow as a function of his distance from the post.

Solution. It is helpful to draw and label a diagram to show what is given and
what is to be found. In the figure below, 24 and 6 correspond to the heights

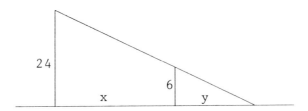

(in feet) of the lamppost and man respectively; x is the distance of the man
from the post, and y is the length of the shadow. The problem asks us to find
a formula that will express y in terms of (as a function of) x. Using similar
triangles, we have

$$\frac{x + y}{24} = \frac{y}{6} .$$

Solving for y,

$$6x + 6y = 24y,$$

$$18y = 6x,$$

$$y = \frac{1}{3} x.$$

If g is the function giving shadow length in terms of distance from the post,
we have

$$g(x) = \frac{1}{3} x.$$

Example 3. A cylindrical can with no top has a surface area of 12π square
inches. Write the total volume of the can as a function of the radius.

Solution. The volume V of a cylindrical can of radius r and height h is

$$V = \pi r^2 h,$$

but here V is a function of two variables r and h. We must eliminate h.
Another condition is given: the surface area (bottom and side) of the can is

99

12π, that is,

$$12\pi = \pi r^2 + 2\pi rh.$$

Solving for h in this equation,

$$h = \frac{12\pi - \pi r^2}{2\pi r},$$

and substituting it into the volume equation, we obtain V as a function of r:

$$V = \pi r^2 \left(\frac{12\pi - \pi r^2}{2\pi r} \right),$$

$$V(r) = \frac{\pi r}{2}(12 - r^2).$$

EXERCISES

1. $V = \frac{1}{3}\pi r^2 h$ and $2r/h = 1/3$. Express V as a function of h.

2. $D = \sqrt{x^2 + y^2}$, $x = 2t$, $y = 3t$. Express D as a function of t.

3. Express the area A of a square as a function of its perimeter P. ($A = s^2$, $P = 4s$.)

4. Express the perimeter of a square as a function of its area.

5. Express the perimeter of a square as a function of its diagonal.

6. Express the area of a circle as a function of its diameter. ($A = \pi r^2$, $d = 2r$.)

7. Express the altitude h of an equilateral triangle as a function of its side length s.

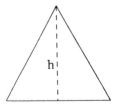

8. Express the area of an equilateral triangle as a function of its side length.

9. A rectangle is inscribed in a circle of radius 5. Express the area of the rectangle as a funcion of x (see figure).

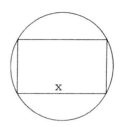

10. A 50 foot rope is threaded around a pulley 25 feet above the ground. As one end of the rope is pulled away, the other end rises (see figure). Express the height y as a function of the distance x.

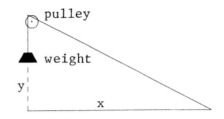

ANSWERS

1. $V = \frac{\pi}{108} h^3$ 2. $d = \sqrt{13}\, t$ 3. $A = P^2/16$ 4. $P = 4\sqrt{A}$

5. $P = 2\sqrt{2}\, d$ 6. $A = \pi d^2/4$ 7. $h = \frac{\sqrt{3}}{2} s$ 8. $A = \frac{\sqrt{3}}{4} s^2$

9. $A = x\sqrt{100 - x^2}$ 10. From the relation $\sqrt{x^2 + 25^2} + (25 - y) = 50$,

we get $y = \sqrt{x^2 + 625} - 25$, valid for $0 \le x \le 25\sqrt{3}$.

24.   Conversion of Units

In this section we list some of the most important relationships between common units of measurement, and indicate how to use them to convert measurements from one set of units to another.

In the metric system, the basic unit of length is the meter and the basic unit of mass is the gram. Under this system, names for greater or smaller measurements are expressed by attaching prefixes. A list of common prefixes is given below, along with the meaning and the standard symbol for each.

Prefix	Symbol	Meaning
mega	M	1000000 times
kilo	k	1000 times
hecto	h	100 times
deca	da	10 times
deci	d	1/10 of
centi	c	1/100 of
milli	m	1/1000 of
micro	μ	1/1000000 of

Thus, a centimeter is 1/100 of a meter, a kilometer is 1000 meters, and so on. The symbol for meter is m and the symbol for gram is g. The prefix symbol is combined with m or g to denote a metric measurement. Thus, 5 cm means 5 centimeters, 7 mg means 7 milligrams, and so on.

In the English system, lengths are measured in units such as inches (in), feet (ft), yards (yd), and miles (mi), and mass is measured in units such as ounces (oz), pounds (lb), and tons. The relationships below hardly need explicit mention.

$$1 \text{ ft} = 12 \text{ in} \qquad 1 \text{ lb} = 16 \text{ oz}$$
$$1 \text{ yd} = 3 \text{ ft} \qquad 1 \text{ ton} = 2000 \text{ lb}$$
$$1 \text{ mi} = 5280 \text{ ft}$$

The conversions between the two systems are defined by the two identites

$$1 \text{ in} = 2.54 \text{ cm}, \quad \text{and} \quad 1 \text{ lb} = 453.59237 \text{ g} \ (\cong 454 \text{ g}).$$

The pound as defined here is a unit of mass. The reader should be warned that the British engineering system uses the pound as a unit of force.

By multiplying each side of these identities by the same constant, you can convert units from the English system to the metric system. To change from the metric system, it is easiest first to rewrite the identities in the form

$$1 \text{ cm} = \frac{1}{2.54} \text{ in,} \quad \text{and} \quad 1 \text{ g} = \frac{1}{454} \text{ lb.}$$

Thus, to convert 1 mile into metric length, start with the identity 1 in = 2.54 cm and multiply each side by 12 to get 1 ft = 30.48 cm. Then, multiply this by 5280 to get 5280 ft = 160,934.4 cm, or equivalently, 1 mi = 1.609344 km. To express 1 m as an equivalent length in the English system, start with 1 cm = 1/2.54 in, and multiply each side by 100 to get 1 m = 39.37007874... in. We have just derived the familiar approximations:

$$1 \text{ mi} \cong 1.6 \text{ km}$$
$$1 \text{ m} \cong 39.37 \text{ in.}$$

Example 1. Express 1500 meters in miles.

Solution. We can string together a series of conversions to get from meters to miles; namely, we can get from meters to inches, from inches to feet, from feet to miles. The relevant formulas are

$$1 \text{ m} = 39.37 \text{ in}$$
$$1 \text{ in} = 1/12 \text{ ft}$$
$$1 \text{ ft} = 1/5280 \text{ mi,}$$

and therefore,

$$1500 \text{ m} = (1500)(39.37)\text{in}$$
$$= (1500)(39.37)(1/12)\text{ft}$$
$$= (1500)(39.37)(1/12)(1/5280)\text{mi}$$
$$\cong .932 \text{ mi.}$$

The conversion formulas for length can be used to derive conversion formulas for measurements of area and volume. For example, one square foot is equal to the product of a one foot length by one foot length, 1 sq ft = (1 ft)·(1 ft), and this is equal to (12 in)(12 in) or 144 sq in; more compactly,

$$1 \text{ ft}^2 = 144 \text{ in}^2.$$

Example 2. Express 5 cubic meters in cubic feet.

Solution. We first find a direct relation between meters and feet. The relevant formulas are

$$1 \text{ m} = 39.37 \text{ in},$$

$$1 \text{ in} = 1/12 \text{ ft},$$

and therefore,

$$1 \text{ m} = (39.37)\text{in} = (39.37/12)\text{ft}.$$

It follows that

$$5\text{m}^3 = 5[1 \text{ m}]^3$$

$$= 5[39.37/12 \text{ ft}]^3$$

$$= 5[39.37/12]^3 \text{ft}^3$$

$$\cong 176.57 \text{ ft}^3.$$

Many physical measurements involve the quotient of two measurements; for example, velocity is a measure of distance per unit time (ft/sec, mi/hr, etc.), density is mass per unit volume (lb/ft^3, kg/m^3, etc.), pressure is force per unit area, and so forth. To convert measurements of this kind from one form to another, we simply work with each of the measurements separately. For example, 60 miles per hour is the same as (60)(5280) ft per hour or (60)(5280) ft per 3600 sec which is equivalent to (60)(5280)/3600 ft per sec. This computation is more conveniently displayed in the following form:

$$60 \text{ mi/hr} = \frac{60 \text{ mi}}{1 \text{ hr}}$$

$$= \frac{(60)(5280)\text{ft}}{3600 \text{ sec}}$$

$$= 88 \text{ ft/sec}.$$

104

Example 3. Express 60 pounds per cubic foot in kilograms per cubic centimeter.

Solution. We need to transform pounds into kilograms, and feet into centi-

meters. The conversion formulas are

$$1 \text{ lb} = 454 \text{ g} = 454/1000 \text{ kg},$$

$$1 \text{ ft} = 12 \text{ in} = (12)(2.54) \text{ cm}.$$

It follows that

$$1 \text{ ft}^3 = (1 \text{ ft})^3 = [(12)(2.54 \text{ cm})]^3 = [(12)(2.54)]^3 \text{cm}^3,$$

and therefore

$$60 \text{ lb/ft}^3 = \frac{60 \text{ lb}}{1 \text{ ft}^3} = \frac{60(\frac{454}{1000}) \text{ kg}}{[(12)(2.54)]^3 \text{cm}^3} \cong .000962 \text{ kg/cm}^3.$$

EXERCISES

Express the measurement on the left as an equivalent measurement in the units

given on the right.

1. 15 km = _____ mi   6. 625 ft^3 = _____ yd^3

2. 440 yd = _____ m   7. 10 ft/sec = _____ mi/hr

3   25 kg = _____ lb   8. 17,000 mi/year = _____ m/min (1 yr = 365 days)

4. 400 mi^2 = _____ km^2 9. 5 g/mm^2 = _____ lb/ft^2

5. 400 ft^2 = _____ m^2 10. 25 kg/m^3 = _____ lb/ft^3

ANSWERS

1. 9.32   2. 402.336   3. 55.1   4. 1036   5. 37.161

6. 23.148   7. 6.8181...   8. 52.05   9. 1024   10. 1.56

25. Algebraic Manipulations

We have seen (Section 8) that linear equations are easy to graph when they are converted to the standard slope-intercept form:

$$y = mx + b.$$

Also, we have seen (Section 17) that to apply the quadratic formula we must first write the quadratic equation in the form

$$ax^2 + bx + c = 0.$$

These are two instances of the fact that the application of a general theory often requires that the equation first be put into a standard form. You will see further examples of this in the next section, and in calculus when you study methods of integration. In this section we will practice putting algebraic expressions into prescribed forms and give applications of the *give and take* principle.

__Example 1.__ Write $x^2 + 3x + 5$ in the form $(x - a)^2 + b^2$, $b > 0$.

__Solution.__ We write the expression in the form

$$(x^2 + 3x + \underline{}) + 5$$

and then "give and take" 9/4 to complete the square within the parentheses:

$$(x^2 + 3x + \frac{9}{4}) + 5 - \frac{9}{4},$$

$$(x + \frac{3}{2})^2 + \frac{11}{4},$$

$$[x - (-\frac{3}{2})]^2 + (\frac{\sqrt{11}}{2})^2.$$

This is the required form: $a = -3/2$, $b = \sqrt{11}/2$.

<u>Example 2.</u> Write $3y + 4x^2 - 7x + 9 = 0$ in the form $y = A(x - h)^2 + k.$

<u>Solution.</u> We first solve for y in the original equation, and get

$$y = -\frac{4}{3} x^2 + \frac{7}{3} x - 3.$$

In this form it is not easy to complete the square in the x-terms, so we

factor $- 4/3$ from each of the x-terms. This can be conveniently carried out

by applying a multiplicative version of the give and take principle: multiply

and divide by $- 4/3$, or equivalently, multiply by $- 4/3$ and $- 3/4$. Thus, we

have

$$y = -\frac{4}{3} [(-\frac{3}{4})(-\frac{4}{3} x^2 + \frac{7}{3} x)] - 3,$$

$$y = -\frac{4}{3} (x^2 - \frac{7}{4} x + \underline{}) - 3.$$

Within the parentheses,   complete the square by adding $[\frac{1}{2}(-\frac{7}{4})]^2$, or $\frac{49}{64}$.

In doing   this, we have really added $-\frac{4}{3}(\frac{49}{64})$, or $-\frac{49}{48}$, to the right side

of the equation, so this must also be taken away:

$$y = -\frac{4}{3} [x^2 - \frac{7}{4} x + \frac{49}{64}] - 3 - (-\frac{49}{48})$$

$$= -\frac{4}{3} (x - \frac{7}{8})^2 + (-\frac{95}{48}).$$

This is the desired form: $A = - 4/3$, $h = 7/8$, $k = - 95/48.$

<u>Example 3.</u> Write the equation $2x + 3y = 4$ in the form $\frac{x}{a} + \frac{y}{b} = 1.$

<u>Solution.</u> An essential feature of the required form is that the constant term

on the right side is 1, so we begin by dividing the original equation by 4:

$$\frac{x}{2} + \frac{3y}{4} = 1.$$

Now give and take by multiplying the numerator and denominator of the y coeffi-

cient by 1/3 :

$$\frac{x}{2} + \frac{y}{4/3} = 1.$$

This is the prescribed form: $a = 2$, $b = 4/3$.

Example 4. Write $16x^2 - 9y^2 - 64x + 18y - 89 = 0$ in the form

$$\frac{(x - h)^2}{a^2} - \frac{(y - k)^2}{b^2} = 1.$$

Solution. The first step is to complete the squares among the x-terms and y-terms respectively. Write the equation in the form

$$(16x^2 - 64x + \underline{\quad}) - (9y^2 - 18y + \underline{\quad}) = 89,$$

$$16(x^2 - 4x + \underline{\quad}) - 9(y^2 - 2y + \underline{\quad}) = 89.$$

Then complete the squares:

$$16(x^2 - 4x + 4) - 9(y^2 - 2y + 1) = 89 + 64 - 9$$

(we've added 64 to each side to complete the square in the x's and subtracted 9 from each side to complete the square in the y's). This gives

$$16(x - 2)^2 - 9(y - 1)^2 = 144.$$

We divide by 144 to get

$$\frac{16(x - 2)^2}{144} - \frac{9(y - 1)^2}{144} = 1,$$

and this can be written as

$$\frac{(x - 2)^2}{3^2} - \frac{(y - 1)^2}{4^2} = 1.$$

This is the desired form: $h = 2$, $k = 1$, $a = 3$, and $b = 4$.

1. Write $x^2 - 4x + 7$ in the form $(x - a)^2 + b$.

2. Write $x^2 + 6x + 4$ in the form $(x + a)^2 - b^2$.

3. Write $7x + 3$ in the form $a(x + b)$.

4. Write $7x^2 + 2x + 3$ in the form $A(x^2 + Bx + C)$.

5. Write $4x^2 + 12x + 7$ in the form $A(x + h)^2 + k$.

6. Write $4x + 2y - 5 = 0$ in the form $y = a(x - b)$.

7. Write $y^2 + 2y - 8x + 25 = 0$ in the form $x = A(y - k)^2 + h$.

8. Write $5x^2 - 3y^2 - 12 = 0$ in the form $\dfrac{x^2}{a^2} - \dfrac{y^2}{b^2} = 1$.

9. Write $9x^2 + 25y^2 - 36x + 150y + 36 = 0$ in the form $\dfrac{(x - h)^2}{a^2} + \dfrac{(y - k)^2}{b^2} = 1$.

10. Write $xy' + y = 2xy + x^2 y' + 3$ in the form $y' + P(x)y = Q(x)$, where $P(x)$ and $Q(x)$ are functions of x.

ANSWERS

1. $(x - 2)^2 + 3$ 2. $(x + 3)^2 - (\sqrt{5})^2$ 3. $7(x + \frac{3}{7})$

4. $7(x^2 + \frac{2}{7}x + \frac{3}{7})$ 5. $4(x + \frac{3}{2})^2 + (-2)$ 6. $-2(x - \frac{5}{4})$

7. $x = \frac{1}{8}(y - (-1))^2 + 3$ 8. $\dfrac{x^2}{\left(\sqrt{\frac{12}{5}}\right)^2} - \dfrac{y^2}{(2)^2} = 1$

9. $\dfrac{(x - 2)^2}{5^2} + \dfrac{(y - (-3))^2}{3^2} = 1$ 10. $y' + \left(\dfrac{1 - 2x}{x(1 - x)}\right)y = \dfrac{3}{x(1 - x)}$

26. Circles, Ellipses, Parabolas, Hyperbolas

Circle

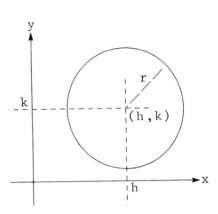

Consider a circle of radius r with center at (h,k). A point P = (x,y) in the plane is on the circle if and only if its distance from the center is r; that is (see Section 9, page 32), if and only if

$$\sqrt{(x - h)^2 + (y - k)^2} = r,$$

or

$$(x - h)^2 + (y - k)^2 = r^2. \qquad (1)$$

It follows that any equation that can be put into the form of equation (1) represents a circle with center (h,k) and radius r.

Example 1. Show that the graph of the equation below is a circle, and find its center and radius: $x^2 - 2x + y^2 + 3y = 2$.

Solution. We write the equation in the form

$$(x^2 - 2x + \underline{\quad}) + (y^2 + 3y + \underline{\quad}) = 2,$$

and add the appropriate constants to each side of the equation to complete the square. In the first case we add 1, and in the second we add 9/4 to get

$$(x^2 - 2x + 1) + (y^2 + 3y + \frac{9}{4}) = 2 + 1 + \frac{9}{4},$$

$$(x - 1)^2 + (y + \frac{3}{2})^2 = \frac{21}{4}.$$

This is the standard form for the equation of a circle of radius $\frac{\sqrt{21}}{2}$ and center (1,-3/2).

Ellipse

The equation

$$\frac{x^2}{r^2} + \frac{y^2}{r^2} = 1$$

represents a circle of radius r with center at the origin. This is a special case of the more general equation

$$\frac{x^2}{a^2} + \frac{y^2}{b^2} = 1, \qquad (2)$$

where a and b are arbitrary numbers. The graph of equation (2) is called an *ellipse*. The x-intercepts are ±a, and the y-intercepts are ±b. The graph is shown below; you can think of an ellipse as a squashed or stretched circle.

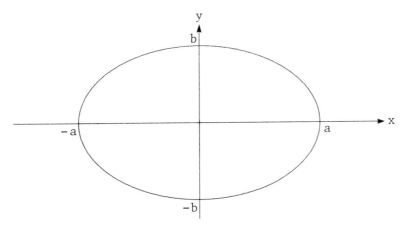

Example 2. Sketch the graph of $16x^2 + 7y^2 = 112$.

Solution. The equation contains both an x^2 term and a y^2 term and it can be put into the standard form for an ellipse (equation (2)). To do this, divide by 112,

$$\frac{16x^2}{112} + \frac{7y^2}{112} = 1,$$

and rewrite this in the form

$$\frac{x^2}{(112/16)} + \frac{y^2}{(112/7)} = 1,$$

$$\frac{x^2}{7} + \frac{y^2}{16} = 1.$$

111

The x-intercepts are $\pm\sqrt{7}$, the y-intercepts are ±4, and the graph is shown below.

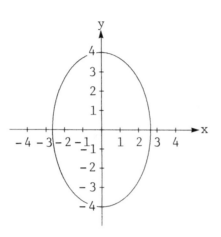

Parabola

The graph of $y = x^2$ is shown at the right;
it is an example of a curve called a *parabola*.
Other examples include all curves whose equations have the form $y = Ax^2$; a few special
cases are shown below.

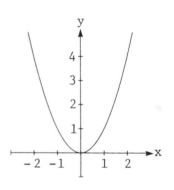

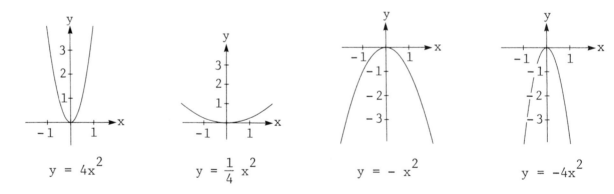

$$y = 4x^2 \qquad y = \frac{1}{4}x^2 \qquad y = -x^2 \qquad y = -4x^2$$

The curves in the figures shown here have several common properties: (i) the
point $(0,0)$ is special in each of them and is called the *vertex* of the parabola; (ii) they are each symmetric about the y-axis which is called the *axis* of
the parabola. Notice that the coefficient A influences the opening of the
parabola. When A is positive, the parabola opens upward, and when A is nega-

112

tive it opens downward. As A decreases in magnitude, the opening of the parabola increases.

The graph of $y = (x - 1)^2 + 2$ is shown at the right. As x moves away from 1, the corresponding value for y will increase because $(x - 1)^2$ is positive, and this increase varies with the square of the distance between x and 1. The graph of $y = (x - 1)^2 + 2$ is also a parabola; its vertex is the point (1,2) and its axis is the vertical line x = 1. It is simply a translation (one unit to the right and two units upward) of the graph of $y = x^2$.

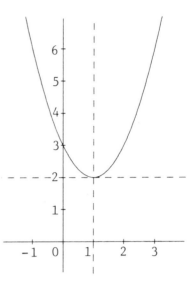

In a similar way, the graph of

$$y = A(x - h)^2 + k \qquad\qquad (3)$$

is a parabola. The vertex is at (h,k) and its axis is the vertical line x = h. The coefficient A controls the direction (upward or downward) and the spread of the opening.

Example 3. Sketch the graph of $3x^2 - 12x + 4y = 0$.

Solution. There is an x^2 term, but no y^2 term. The equation can be put into the standard form of equation (3). To do this, rewrite the original equation in the form

$$4y = -3x^2 + 12x$$
$$= -3(x^2 - 4x + \underline{\quad})$$
$$= -3(x^2 - 4x + 4) + 12$$
$$= -3(x - 2)^2 + 12,$$

and divide each side by 4 to get

113

$$y = -\frac{3}{4}(x - 2)^2 + 3.$$

In this form, we recognize (2,3) as the vertex of the parabola; the value of y decreases as x moves away from 2 because of the negative coefficient, $-\frac{3}{4}$. To estimate the spread, we merely plot a couple of points: when x = 0 we get y = 0 and when x = 4 we get y = 0. The sketch is shown below.

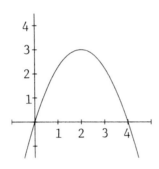

The graph of $x = y^2$ shown at the right is a parabola with a horizontal axis. Each positive x corresponds to two y-values:

$$y = \pm\sqrt{x}.$$

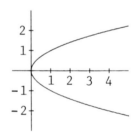

More generally, any equation of the form

$$x = A(y - k)^2 + h \qquad\qquad (4)$$

represents a parabola with vertex at (h,k) and with   horizontal axis, y = k.

Hyperbola

Consider the graph of the equation xy = 1. Write the equation in the form

$$y = \frac{1}{x}$$

and think of y as a function of x. Notice that y varies inversely with x. As x *increases* from 1 to infinity, y *decreases* from 1 to 0; as x *decreases* from 1 to 0, y *increases* from 1 to infinity. Similarly, as x decreases from -1 to minus infinity, y increases from -1 to 0; as x increases from -1 to 0, y decreases from -1 to minus infinity. The graph is shown below.

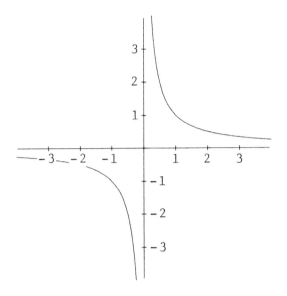

This curve is called a *hyperbola*. The x and y axes are *asymptotes*; the graph of the hyperbola gets closer and closer to these lines as it is traced farther and farther away from the origin.

The equation xy = 1 is a special case of the more general equation

$$(x - a)(y - b) = A. \qquad\qquad (5)$$

Any equation that can be put into this form also describes a hyperbola. We can make an analysis of this equation, in the same way we did for xy = 1, by rewriting it in the form

$$y = b + \frac{A}{x - a} .$$

As x approaches infinity, y approaches b, and as x approaches a, y becomes large in magnitude. A sketch of the general equation (5) has the shape shown in the next figure (when A is positive). The asymptotes are the lines x = a and y = b.

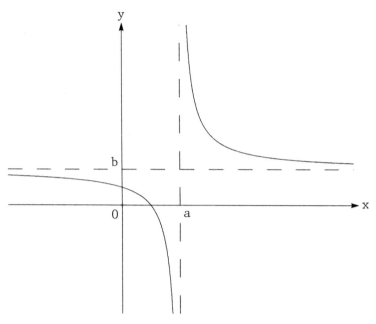

Another class of hyperbolas is defined by equations of the form

$$\frac{x^2}{a^2} - \frac{y^2}{b^2} = 1.\qquad(6)$$

We may assume that a and b are positive. The graph can be visualized most easily after solving the equation for y:

$$\frac{y^2}{b^2} = \frac{x^2}{a^2} - 1,$$

$$y^2 = \frac{b^2}{a^2}(x^2 - a^2)$$

$$y = \pm\frac{b}{a}\sqrt{x^2 - a^2}\,.$$

For y to be defined, it is necessary that $|x| > |a|$ ($\sqrt{x^2 - a^2}$ is imaginary when $|x| < |a|$). The x-intercepts are $\pm a$. The value for $|y|$ increases as $|x|$ increases. When $|x|$ is *very* large, a^2 will be insignificant compared to x^2, and therefore $\sqrt{x^2 - a^2} \cong \sqrt{x^2}$. For such x, $y \cong \pm\frac{b}{a}x$. This rough analysis indicates that the hyperbolas given by equation (6) have the general shape shown in the next figure. The lines $y = \frac{b}{a}x$ and $y = -\frac{b}{a}x$ are asymptotes.

116

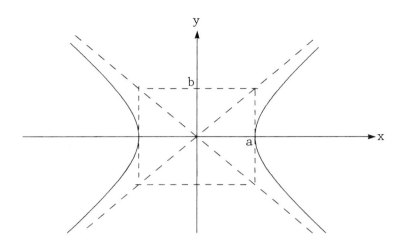

In a similar manner, the difference of squares,

$$\frac{y^2}{b^2} - \frac{x^2}{a^2} = 1, \qquad (7)$$

defines a family of hyperbolas with asymptotes $y = \pm \frac{b}{a} x$. (Solve for y: $y = \pm \frac{b}{a} \sqrt{x^2 + a^2}$.) Their graphs have the form sketched below.

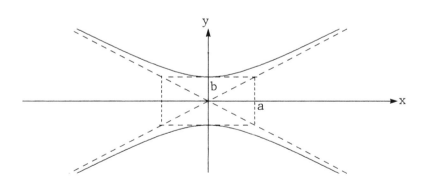

Example 4. Sketch the graph of $9y^2 - 4x^2 = 36$.

Solution. The equation has the standard form (7) of a hyperbola:

$$\frac{y^2}{4} - \frac{x^2}{9} = 1.$$

Solving for y, we have

117

$$\frac{y^2}{4} = \frac{x^2}{9} + 1$$

$$y = \pm \frac{2}{3} \sqrt{x^2 + 9} \ .$$

The graph is shown below.

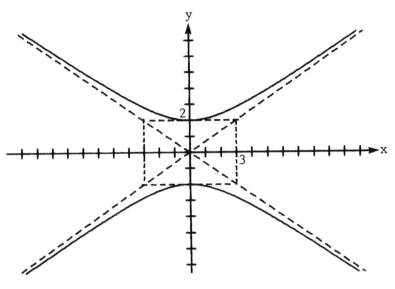

We conclude this section with three remarks.

(i) The curves studied in this section are called the *conic* sections. This is because they are the curves you get when you intersect a cone (a double-napped cone in the case of a hyperbola) and a plane tilted at various angles.

(ii) The emphasis in this section has been on graphing. Notice that each of the equations considered has the general form

$$Ax^2 + Bxy + Cy^2 + Dx + Ey + F = 0.$$

(iii) The study of the conic sections has been carried on since the Greeks of antiquity. The subject can be approached geometrically (as conic sections), or algebraically (as equations of the type given in (ii)), or a combination of these. A third approach is to define each of the conics as a "locus" (set) of points satisfying a certain geometric property.

118

EXERCISES

Identify the conic sections below and sketch the graph after putting the equation in the appropriate standard form (equations (1)-(7)).

1. $x^2 + 4x + y^2 - 6y = 3$

2. $(x - 1)^2 + (y - 3)^2 = (2x + 1)^2 + (2y + 1)^2$ (Begin by expanding.)

3. $4x^2 + 9y^2 = 36$ 7. $x = -4y^2$

4. $4x^2 + y^2 = 1$ 8. $y = \dfrac{-2}{x - 1}$

5. $y = x^2 - 2$ 9. $9x^2 - 16y^2 = 144$

6. $y = -x^2 + 2x + 3$ 10. $y^2 - x^2 = 4$

ANSWERS

1.

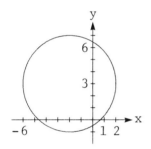

Center: $(-2, 3)$

Radius: 4

2.

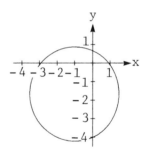

Center: $(-1, -5/3)$

Radius: $\sqrt{58/3}$

3. $\dfrac{x^2}{9} + \dfrac{y^2}{4} = 1$ 4. $\dfrac{x^2}{(1/2)^2} + \dfrac{y^2}{1^2} = 1$

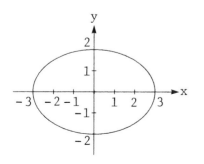

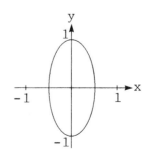

5.

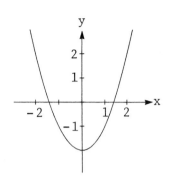

6. $y = -(x - 1)^2 + 4$

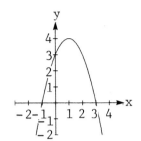

7.

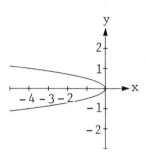

8.

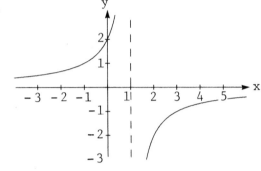

9. $\dfrac{x^2}{4^2} - \dfrac{y^2}{3^2} = 1$

10. $\dfrac{y^2}{2^2} - \dfrac{x^2}{2^2} = 1$

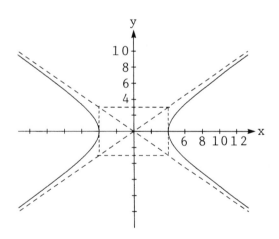

PART IV: Logarithms, Trigonometry

Logarithmic and trigonometric functions are very important in the study of calculus and its applications. For a positive number a, we define $\log_a y$ to be the exponent to which a must be raised to give y. The first two sections explore the meaning and consequences of this definition. This material is not easy, and you may need to study it several times. For the trigonometric sections, concentrate first on learning the definitions of the six trigonometric functions (Sections 30 and 31), particularly the sine and cosine functions. The early mastery of this material will make the succeeding sections much easier. You will need to solve trigonometric equations (Section 33) when you study applications of calculus that require the use of trigonometric functions. Trigonometric identities (Section 34) and inverse trigonometric functions (Section 35) will be used when you study methods of integration. The final section reviews the elementary arithmetic and geometry of complex numbers.

DIAGNOSTIC TEST: Over Part IV

(Answers on page 188)

1. Solve for x: $\log_3(x + 1) = 2$.

2. Write $\frac{1}{2} \log_a(x + 2) - 3 \log_a(x - 1) + 2 \log_a x$ as a single logarithm.

3. Solve for x:

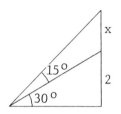

4. Express $\sec \frac{\pi}{4} + \tan \frac{\pi}{6}$ in terms of radicals.

5. Which (if any) of the following equations are valid for all θ?

   (a) $\sin \theta = - \sin \theta$ (b) $\sin(\frac{\pi}{2} - \theta) = \cos \theta$ (c) $\cos(\pi - \theta) = \cos \theta$

6. Sketch the graph of $y = 2 \sin 2\theta$.

7. Find all θ between 0 and 2π which satisfy $\cos 3\theta = 0$.

8. Express $\tan \theta [\frac{1}{\csc \theta} + \frac{\cot \theta}{\sec \theta}]$ in terms of sines and cosines and simplify.

9. Which (if any) of the following equations are valid for all θ?

   (a) $\cos 2\theta = \cos^2 \theta - \sin^2 \theta$ (b) $\sin^2(\frac{\theta}{2}) = \frac{1 - \sin \theta}{2}$

   (c) $\cos 3\theta = \sin \theta \cos 2\theta - \sin 2\theta \cos \theta$

10. Evaluate $\sin(\text{Arccos } x)$, $0 \le x \le 1$.

11. What is the magnitude and argument of $(-1 + \sqrt{3} \; i)^2$?

27. Logarithms, Definition

Graphs of the exponential function $f(x) = a^x$ for various values of (positive) a are shown below.

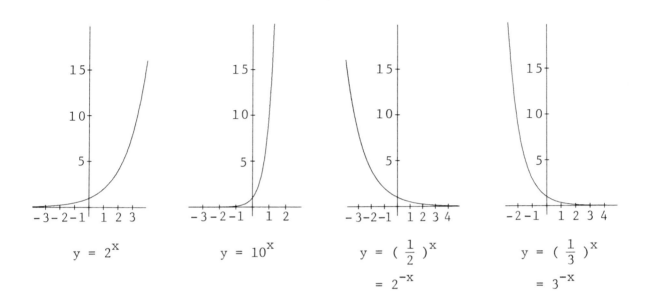

$$y = 2^x \qquad y = 10^x \qquad y = (\tfrac{1}{2})^x \qquad y = (\tfrac{1}{3})^x$$
$$= 2^{-x} \qquad\quad = 3^{-x}$$

Previously, in these notes, a^x was defined only for rational values of x (Section 14). This definition can be extended in exactly one way to all values of x so that the graph of $y = a^x$ is smooth and continuous. Without going into details, we will assume this extended definition is being used. The rules of exponents given in Section 14 continue to hold for all *real* exponents. We will only be concerned with exponential functions a^x for positive values of a.

The graphs in the preceding figure suggest the following features for exponential functions. First note that $a^x > 0$ for *all* values of x. Furthermore, if $a > 1$, then a^x increases without bound as x tends toward infinity. This is symbolized by writing $a^x \to \infty$ as $x \to \infty$. Also, for $a > 1$, $a^x \to 0$ as $x \to -\infty$ (a^x gets closer and closer to zero as x tends toward minus infinity, that is, as x becomes more and more negative without bound). Similarly, for $0 < a < 1$, we have $a^x \to 0$ as $x \to \infty$ and $a^x \to \infty$ as $x \to -\infty$. (You should check these tendencies in the four graphs shown in the preceding figure.)

Consider the exponential function $y = a^x$, where a is a positive number, $a \neq 1$. If y is any positive number there is exactly one value x for which

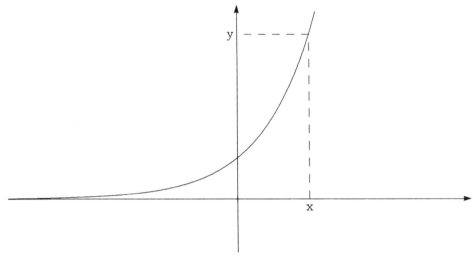

$a^x = y$ (see the preceding graph). This number x is called the logarithm of y base a, and is written $\log_a y$. Restated in words, $\log_a y$ *is the exponent to which a must be raised to give y.* Thus, from the definition,

$$(i) \qquad a^{\log_a y} = y, \quad y > 0.$$

A particularly useful formulation of the definition is:

$$x = \log_a y \quad \text{if and only if} \quad a^x = y.$$

You should work out some way of remembering the mechanics of this definition because changing an equation from logarithmic form to an equivalent equation in exponential form, and vice versa, is the key to solving many logarithmic and exponential problems.

Example 1. a) Find the value of $\log_3 81$. b) Solve for x in the equation $\log_5 x = 4$. c) Solve for x in the equation $\log_x 3 = 1/5$. d) Solve for x in

124

the equation $\log_{10}(x + 3) = 2y + 7$.

Solution. a) Set $\log_3 81 = y$. Then $3^y = 81$, and it follows that $y = 4$.

b) The equation $\log_5 x = 4$ is equivalent to $5^4 = x$, or $x = 625$.

c) The equation $\log_x 3 = 1/5$ is equivalent to $x^{1/5} = 3$. Raising each side to the fifth power gives $x = 3^5 = 243$.

d) $x = 10^{2y+7} - 3$.

Return now to a consideration of the preceding figure, the graph of $y = a^x$. We can think of the positive y-axis as the *domain* of a function f which associates each positive number y with its logarithm $\log_a y$; symbolically, $f(y) = \log_a y$. To draw a graph of this function, it is standard to place the domain of the function along the horizontal axis, and to let x denote the independent variable (exchange axes and relabel). When we do this we get the graph $(f(x) = \log_a x)$ below.

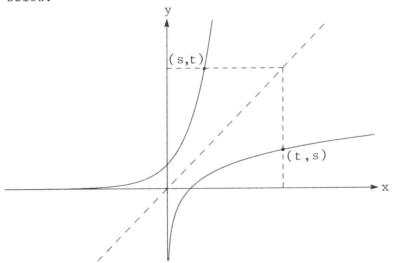

The dotted graph is that of $y = a^x$. The two graphs are reflections of one another about the line $y = x$. If this is not clear from the geometric description, a formal proof follows from the fact that a point (s,t) is on the graph of $y = a^x$ if and only if (t,s) is on the graph of $y = \log_a x$. [(s,t) is on

$y = a^x$ if and only if $t = a^s$; if and only if $s = \log_a t$; if and only if (t,s) is on the graph of $y = \log_a x$.]

Example 2. Sketch the graph of $y = \log_2 x$.

Solution. We may either reflect the graph of $y = 2^x$ about the line $y = x$, or we may plot points by making a table.

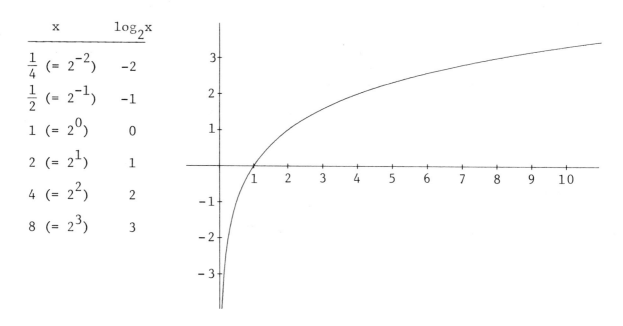

x	$\log_2 x$
$\frac{1}{4}$ $(= 2^{-2})$	-2
$\frac{1}{2}$ $(= 2^{-1})$	-1
1 $(= 2^0)$	0
2 $(= 2^1)$	1
4 $(= 2^2)$	2
8 $(= 2^3)$	3

There are certain logarithmic values that should be memorized because they occur so frequently. In addition to (i) given earlier, these include

$$\text{(ii)} \qquad \log_a 1 = 0,$$

$$\text{(iii)} \qquad \log_a a = 1,$$

$$\text{(iv)} \qquad \log_a a^x = x.$$

These are proved directly from the definition. For example, for (ii), set $\log_a 1 = y$. Then $a^y = 1$, so that $y = 0$. For (iii), set $\log_a a = y$. It follows that $a^y = a$, or $y = 1$. For (iv), set $\log_a a^x = y$. Then $a^y = a^x$, or $y = x$. These proofs are typical of those used in proving logarithmic identities: (1) set the logarithms in the formula equal to variables, (2) change into exponential form, and (3) use the properties of exponents to finish the problem.

126

This standard technique is very useful; we will use it again in the next section.

EXERCISES

1. Rewrite each of the following equalities in an equivalent exponential form.

   a) $\log_3 27 = 3$ b) $\log_2 \sqrt{2} = 1/2$ c) $\log_{10} 100 = 2$

2. Use the definition of the logarithm (as opposed to a hand calculator) to evaluate the following logarithms.

   a) $\log_2 32$ b) $\log_{10} 100,000$ c) $\log_3 1/27$

3. Solve for x in the following equations (change to the equivalent exponential form).

   a) $\log_2 (1 + x) = 3$ b) $\log_5 (1/x) = 1$ c) $\log_{10} \sqrt{x} = 6$

4. Simplify the following expressions.

   a) $\log_{10} 10^x$ b) $5^{\log_5 x}$ c) $2^{\log_2 x^2}$

5. Solve for x in the following equations.

   a) $\log_a 2x = 1$ b) $\log_a (3 + x) = 1$ c) $\log_a (2x - 2) = 0$

6. Solve for x in the following equation: $\log_x (5x - 6) = 2$.

7. Solve for y (in terms of x) in each of the following equations.

   a) $\log_{10} \sqrt{y} = x^2$ b) $\log_a y = kx + C$ c) $\log_a (2y - 3) = x$

For each of the following problems, sketch the three given equations on the same coordinate axes.

8. a) $y = \log_2 x$

x	1/4	1/2	1	2	4
y					

b) $y = \log_3 x$

x	1/9	1/3	1	3	9
y					

c) $y = \log_5 x$

x	1/25	1/5	1	5	25
y					

9. a) $y = \log_2 x$

x	1/4	1/2	1	2	4
y					

b) $y = \log_2 2x$

x	1/4	1/2	1	2	4	8
y						

c) $y = \log_2 3x$

x	1/6	1/3	2/3	4/3	8/3	16/3
y						

10. a) $y = \log_2 (1 + x)$ b) $y = \log_2 (4 + x)$ c) $y = \log_2 (9 + x)$

ANSWERS

1. a) $3^3 = 27$,   b) $2^{1/2} = \sqrt{2}$,   c) $10^2 = 100$ 2. a) 5,   b) 5,   c) -3

3. a) 7,   b) 1/5,   c) 10^{12} 4. a) x,   b) x,   c) x^2

5. a) $x = a/2$,   b) $x = a - 3$,   c) $x = 3/2$ 6. $x = 2$, $x = 3$

7. a) $y = 10^{2x^2}$,   b) $y = Aa^{kx}$, where $A = a^C$,   c) $y = \dfrac{a^x + 3}{2}$

8.

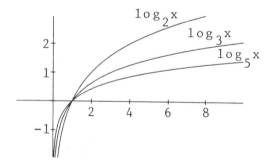

9.

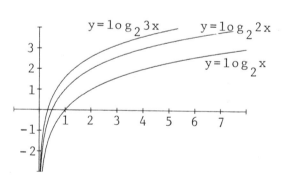

10.

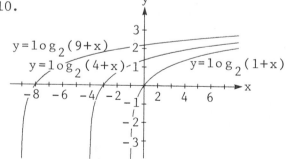

28. Properties of Logarithms

The following properties of logarithms are most important:

$$\text{I.} \quad \log_a xy = \log_a x + \log_a y$$

$$\text{II.} \quad \log_a \left(\frac{x}{y} \right) = \log_a x - \log_a y$$

$$\text{III.} \quad \log_a x^y = y \log_a x$$

These are proved in the manner outlined in the last paragraph of the preceding section. We will illustrate it again by proving the first property and will leave the second and third properties as exercises. For the first property, set $A = \log_a xy$, $B = \log_a x$, and $C = \log_a y$, and change each of these into exponential form to get $a^A = xy$, $a^B = x$, and $a^C = y$. It follows that $a^A = xy = a^B a^C = a^{B+C}$ and therefore $A = B + C$, which is precisely the statement of property I.

Example 1. Solve for y: $\log_a y = \log_a (x - 1) + 2 \log_a (x - 2) - 3 \log_a (x - 4)$.

Solution. Properties I, II, and III can be used to combine sums and differences of logarithms into a single term. For example, we can simplify the right side of the given equation as follows:

$$\log_a y = \log_a (x - 1) + \log_a (x - 2)^2 - \log_a (x - 4)^3, \qquad \text{by III,}$$

$$= \log_a (x - 1)(x - 2)^2 - \log_a (x - 4)^3, \qquad \text{by I,}$$

$$= \log_a \frac{(x - 1)(x - 2)^2}{(x - 4)^3}, \qquad \text{by II.}$$

It follows (see Exercise 5) that

$$y = \frac{(x - 1)(x - 2)^2}{(x - 4)^3}$$

Logarithms make it possible to solve equations having variables in the exponent as shown in the next two examples.

Example 2. Solve for x: $3^{2x+3} \cdot 5^{x+1} = 2^{3x-1}$.

Solution. By taking the logarithm of each side, it is possible to get the variable out of the exponent. We will take the logarithm base 10, because these are easily evaluated with a hand calculator. Thus,

$$\log_{10}(3^{2x+3} \cdot 5^{x+1}) = \log_{10} 2^{3x-1}.$$

Using property I, we expand the left side,

$$\log_{10} 3^{2x+3} + \log_{10} 5^{x+1} = \log_{10} 2^{3x-1},$$

and now, using property III (the point!),

$$(2x + 3)\log_{10} 3 + (x + 1)\log_{10} 5 = (3x - 1)\log_{10} 2.$$

Expanding and separating constant and variable terms, we have

$$(2 \log_{10} 3 + \log_{10} 5 - 3 \log_{10} 2)x = -\log_{10} 2 - 3 \log_{10} 3 - \log_{10} 5$$

$$(\log_{10} 3^2 + \log_{10} 5 - \log_{10} 2^3)x = -(\log_{10} 2 + \log_{10} 3^3 + \log_{10} 5)$$

$$[\log_{10}(\frac{3^2 \cdot 5}{2^3})] \; x = -[\log_{10} 2 \cdot 3^3 \cdot 5]$$

$$x = -\frac{\log_{10}(270)}{\log_{10}(\frac{45}{8})}$$

With the aid of a hand calculator we get

$$x \cong -\frac{2.4313638}{.75012253} \cong -3.2412888.$$

Example 3. Solve for t: $I = \frac{E}{R}(1 - 10^{-(Rt/L)})$.

Solution. Multiplying both sides by R and dividing by E, we get

$$\frac{IR}{E} = 1 - 10^{-(\frac{Rt}{L})},$$

$$10^{-(\frac{Rt}{L})} = 1 - \frac{IR}{E}.$$

Changing this to logarithmic form we have

$$-\left(\frac{Rt}{L}\right) = \log_{10}(1 - \frac{IR}{E}),$$

and solving for t we get

$$t = -\frac{L}{R} \log_{10}(1 - \frac{IR}{E}).$$

EXERCISES

The starred (*) exercises are optional; however, they are instructive and you really should try them. They are included to give you practice at handling the standard proof technique used when working with logarithms (see the last paragraph of Section 27). If you get stuck, read and study the solution, and go on to the next. Return to try them again at a later time, as often as necessary, until the relevant manipulations are second nature.

1. Use properties I, II, and III to write the following expressions as a single logarithm.

 a) $\log_a(x + 1) + \log_a(x - 2) + 2 \log_a(x - 3)$

 b) $\log_a(x - 4) - \log_a(x + 5) + 4 \log_a x$

 c) $\frac{1}{2} \log_a(x + 1) - \frac{1}{2} \log_a(x - 1)$

2. Let $a = \log_{10}2$, $b = \log_{10}3$, and $c = \log_{10}5$. In each of the following equations, express x in terms of a, b, and c.

 a) $x = \log_{10}360$ b) $x = \log_{10}\frac{54}{25}$

 c) $2^x = 3$ (Hint: Take $\log_{10}$ of each side.)

3. Write each of the following logarithms as sums and differences of logarithms that contain only first powers of x.

a) $\log_a \left(\dfrac{x + 1}{x + 2} \right)$ b) $\log_a \sqrt{3x + 1}$ c) $\log_a \dfrac{(x - 1)^2 (2x + 1)^3}{\sqrt[3]{(4x - 1)^2}}$

4.* Prove properties II and III.

5.* Prove that if $\log_a x = \log_a y$ then $x = y$.

6.* Prove the following identity which is useful in calculus:

$$x^y = a^{y \log_a x}.$$

7.* Prove the following identity which can be used to change a logarithm from base b to base a:

$$\log_a x = \dfrac{\log_b x}{\log_b a}.$$

(Hint: Let $y = \log_a x$; change to exponential form, and then take $\log_b$ of each side.)

8.* In the equation $x^x = a^y$, solve for y in terms of x by taking $\log_a$ of each side. Compare the result with Exercise 6.

9.* Which of the following are true? If true, give a proof; if false, give a counterexample.

a) $\log_a \left(\dfrac{x}{y} \right) = \dfrac{\log_a x}{y}$ b) $\log_a \left(\dfrac{x}{y} \right) = \dfrac{\log_a x}{\log_a y}$ c) $\log_{\frac{1}{a}} x = \log_a \left(\dfrac{1}{x} \right)$

10. a) Solve for y: $3 \log_a y = 2 \log_a x + \log_a C$

b) Solve for R: $Q = CE(1 - 10^{-t/(CR)})$

132

1. a) $\log_a[(x + 1)(x - 2)(x - 3)^2]$ b) $\log_a[\frac{(x - 4)x^4}{x + 5}]$ c) $\log_a\sqrt{\frac{x + 1}{x - 1}}$

2. a) $3a + 2b + c$ b) $a + 3b - 2c$ c) b/a

3. a) $\log_a(x + 1) - \log_a(x + 2)$ b) $\frac{1}{2}\log_a(3x + 1)$

c) $2\log_a(x - 1) + 3\log_a(2x + 1) - \frac{2}{3}\log_a(4x - 1)$

4. Property II: Let $A = \log_a(x/y)$, $B = \log_a x$, $C = \log_a y$. Then $a^A = x/y = a^B/a^C = a^{B-C}$, so $A = B - C$.

Property III: Let $A = \log_a x^y$, $B = \log_a x$. Then $a^A = x^y = (a^B)^y = a^{yB}$, so $A = yB$.

5. Let $A = \log_a x$, $B = \log_a y$. Then $x = a^A$, and $y = a^B$. But $A = B$, and therefore $a^A = a^B$, which is the same as $x = y$.

6. $a^{y\log_a x} = a^{\log_a x^y}$, using property III. By (i) of Section 27, $a^{\log_a x^y} = x^y$.

Alternatively, set $A = a^{y\log_a x}$. Then $\log_a A = \log_a a^{y\log_a x} = y\log_a x \log_a a = y\log_a x = \log_a x^y$. Applying the result of Exercise 5 to the equation $\log_a A = \log_a x^y$ we get $A = x^y$.

7. Let $y = \log_a x$. Then $x = a^y$, and therefore $\log_b x = \log_b a^y = y\log_b a$. Comparing the ends of this equation and solving for y gives

$y = (\log_b x)/(\log_b a)$.

8. $x\log_a x = y\log_a a = y$ 9. a) FALSE. Take $a = 10$, $x = 1$, $y = 10$; $\log_{10}(1/10) = -1$, whereas $(\log_{10}1)/10 = 0$ b) FALSE. Take $a = 10$, $x = 1$, $y = 10$; $\log_{10}(1/10) = -1$, whereas $(\log_{10}1)/(\log_{10}10) = 0$ c) TRUE. Set $y = \log_{1/a}x$. Then $(1/a)^y = x$; $1/a^y = x$; $1/x = a^y$; $y = \log_a(1/x)$.

10. a) $y = \sqrt[3]{Cx^2}$ b) $R = \dfrac{-t}{C\log_{10}(1 - \frac{Q}{CE})}$.

29. Angle Measurement

Angles are measured in degrees and in radians. Measurement in degrees is based on dividing the circumference of a circle into 360 equal parts; this method of measurement is probably most familiar to the reader. However, in calculus, radian measure is preferred because it makes the derivatives of the trigonometric functions simpler. Suppose that an angle is to be measured (see figure). Construct a *unit* circle (radius 1) with center at the vertex of the angle. The *radian measure* of the angle BAC is defined to be the length of the circular arc BC which is subtended by angle BAC.

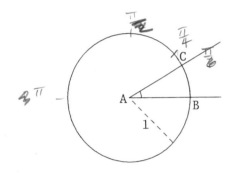

The radian measure of ∠BAC is the length of the arc BC.

The fundamental identity which relates degrees and radians is given by the identity

$$2\pi \text{ radians} = 360 \text{ degrees.}$$

This identity follows from the fact that an angle corresponding to a complete circle equals 360 degrees in degree measure and 2π radians in radian measure (the circumference of the unit circle for such an angle has length 2π).

Example 1. Convert to radians: 180°, 90°, 30°, 45°.

Solution. 180° corresponds to one-half of a circle, so the corresponding radian measure is $\frac{1}{2}$ (2π) = π. 90° corresponds to $\frac{1}{4}$ (2π) = $\pi/2$ radians, 30°

134

is 1/3 of 90°, or $\frac{1}{3}$ ($\pi/2$) = $\pi/6$ radians, and 45° = $\frac{1}{2}$ (90°), so its radian measure is $\frac{1}{2}$ ($\pi/2$) = $\pi/4$.

Example 2. Convert from radians to degrees: 3π, $\pi/3$, $3\pi/4$.

Solution. Since π radians corresponds to 180°, 3π corresponds to 540°; $\pi/3$ is 60°; and $3\pi/4$ is $\frac{3}{4}$ (180°) = 135°.

Example 3. a) Convert 73° to radians.

   b) Convert $\pi/7$ radians to degrees.

Solution. It is probably easiest to begin with the identity 2π radians = 360 degrees. To find 73 degrees in terms of radians, divide by 360 (to get 1 degree in terms of radians) and multiply by 73 to get

$$73 \text{ degrees} = (\frac{2\pi}{360} \cdot 73) \text{ radians}$$

$$\cong 1.2740904 \text{ radians.}$$

To find $\pi/7$ radians, divide the fundamental identity, 2π radians = 360 degrees, by 2π (to get 1 radian in terms of degrees (= 57.29578 degrees)), and multiply by $\pi/7$ to get

$$\pi/7 \text{ radians} = (\pi/7)(360/2\pi) \text{ degrees}$$

$$\cong 25.714286 \text{ degrees.}$$

EXERCISES

Convert from degrees to radians.

1. 60°   2. −45°   3. 225°   4. 150°   5. 36°

Convert from radians to degrees.

6. $\pi/2$ 7. $3\pi/2$ 8. $5\pi/4$ 9. $\frac{5}{3}\pi$ 10. $4\pi/5$

ANSWERS

1. $\pi/3$ 2. $-\pi/4$ 3. $5\pi/4$ 4. $5\pi/6$ 5. $\pi/5$

6. $90°$ 7. $270°$ 8. $225°$ 9. $300°$ 10. $144°$

---

30.   Right Angle Trigonometry

If the two right triangles below are such that $\angle A = \angle A'$, then the two triangles are similar and the ratios of corresponding sides are equal:

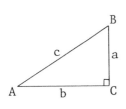
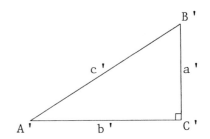

$\frac{a}{c} = \frac{a'}{c'}$, $\frac{b}{c} = \frac{b'}{c'}$, etc. The ratios of the sides do not depend upon the size of the triangle, but are completely determined by the angle A. There are six ratios that can be formed from the three sides of a triangle; the values of these ratios are functions of the angle A. These functions, the sine, cosine, tangent, cosecant, secant, and cotangent, are defined as follows (using the usual abbreviations):

$$\sin A = \frac{a}{c} = \frac{\text{opposite}}{\text{hypotenuse}} \qquad \csc A = \frac{c}{a} = \frac{\text{hypotenuse}}{\text{opposite}}$$

$$\cos A = \frac{b}{c} = \frac{\text{adjacent}}{\text{hypotenuse}} \qquad \sec A = \frac{c}{b} = \frac{\text{hypotenuse}}{\text{adjacent}}$$

$$\tan A = \frac{a}{b} = \frac{\text{opposite}}{\text{adjacent}} \qquad \cot A = \frac{b}{a} = \frac{\text{adjacent}}{\text{opposite}}$$

136

The functions in the right column are the reciprocals of the corresponding functions (same row) in the left column.

Two important right triangles which occur again and again in trigonometric problems are the 45°-triangle in which the two legs are of equal length, and

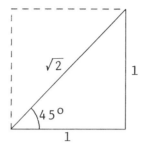

the 30°-60°-90° triangle in which the hypotenuse is twice as long as the shortest leg. (The relationships of the sides can be derived and recalled by dissecting a square and equilateral triangle as shown in the figures.) By refer-

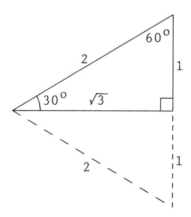

ring to these triangles, it is possible to evaluate the trigonometric functions for A = 30°, 45°, and 60° (respectively, π/6, π/4, and π/3 radians). Their values need not be memorized; however, because they occur so frequently, you should be able to find them by drawing the appropriate triangle. For example,

$\cos 45° = \dfrac{\text{adjacent}}{\text{hypotenuse}} = 1/\sqrt{2} = \sqrt{2}/2, \quad \sin 60° = \dfrac{\text{opposite}}{\text{hypotenuse}} = \sqrt{3}/2, \quad \tan 30° =$

$\dfrac{\text{opposite}}{\text{adjacent}} = 1/\sqrt{3} = \sqrt{3}/3.$

Values for the trigonometric functions of other angles can be found in tables or with a hand calculator. If using a hand calculator, be sure to take into account the manner in which you enter the angles: degrees or radians. The following values (eight-digit approximations) for angles of 41 degrees and $\pi/7$ radians were obtained on a hand calculator.

$$\sin 41° = .65605903 \qquad \sin \pi/7 = .43388374$$

$$\cos 41° = .75470958 \qquad \cos \pi/7 = .90096887$$

$$\tan 41° = .86928673 \qquad \tan \pi/7 = .48157462$$

If the angles and one side of a triangle are known, then the other sides are uniquely determined. Suppose in the right triangle ABC shown below that angle A is known (angle C is a right angle), and let c denote the length of the hypotenuse; let x and y denote the lengths of the legs (see figure). Then

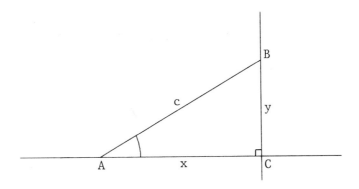

$$\frac{x}{c} = \cos A, \quad \text{or equivalently,} \quad x = c \cos A,$$

and

$$\frac{y}{c} = \sin A, \quad \text{or equivalently,} \quad y = c \sin A.$$

Geometrically, x is the *projection* of the hypotenuse onto the horizontal line AC. This length is equal to that of the hypotenuse c reduced by the factor cos A. Similarly, y is the length of the *projection* of the hypotenuse onto the vertical line CB; it equals the length of the hypotenuse reduced by the factor sin A.

In the figure below suppose that angle BAC = 41° and that AC = 300 feet.

138

Denote the lengths of the other sides by x and y (see figure). Then

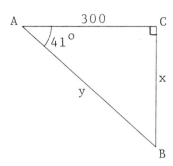

$$\frac{x}{300} = \tan 41°, \quad \text{so that} \quad x = 300 \tan 41° = 260.78602 \text{ feet},$$

and

$$\frac{300}{y} = \cos 41°, \quad \text{so that} \quad y = \frac{300}{\cos 41°} = 397.5039 \text{ feet}.$$

Trigonometry (of right triangles) can be applied in very striking ways, especially in surveying-like problems.

Example 1. A man walks along the bank of a straight river. At a certain point the angle between the bank of the river and a certain tree on the opposite bank is 38 degrees. He walks 100 meters farther along the bank and then finds that the angle between the bank and the same tree is 46 degrees. How wide is the river?

Solution. As in all verbal problems of this sort, the first step is to draw a diagram and label it. For this problem, we have the following diagram:

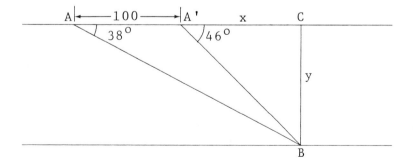

The tree is located at B; when the man is at A, the angle between the bank and the tree is 38°, and at A' (100 meters away) the angle is 46°. We wish to find y, the width of the river. If the length from A to C were known we could do the problem because then we would know a side and the angles of ΔABC. However, in this problem we are not given the length of any of the sides of either ΔABC or ΔA'BC. The trick is to introduce another variable: let x denote the distance from A' to C. Then in ΔA'BC,

$$\tan 46° = \frac{y}{x} ,$$

and in ΔABC,

$$\tan 38° = \frac{y}{x + 100} .$$

We now have two equations and two unknowns which we can solve simultaneously. From the first equation, $y = x \tan 46°$. Substituting this into the second equation, we get

$$\tan 38° = \frac{x \tan 46°}{x + 100} ,$$

$$(x + 100) \tan 38° = x \tan 46°.$$

$$(\tan 46° - \tan 38°) x = 100 \tan 38°,$$

$$x = \frac{100 \tan 38°}{\tan 46° - \tan 38°}$$

which, with the help of a hand calculator, equals (approximately) 307.29673. The width of the river is therefore $(307.29673) \tan 46° \cong 318.21508$ meters.

Example 2. Express the area of the trapezoid below as a function of the angle θ.

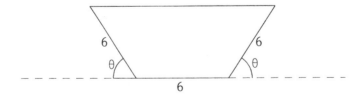

Solution. The area of the trapezoid is $A = \frac{1}{2} (b_1 + b_2)h$. As an intermediate step, we introduce auxiliary variables x and y as shown below.

140

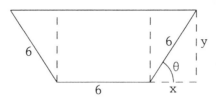

Then we have

$$A = \frac{1}{2} [6 + (6 + 2x)]y,$$

$$A = (6 + x)y.$$

But $x = 6 \cos \theta$ (the horizontal projection of the hypotenuse) and $y = 6 \sin \theta$ (the vertical projection of the hypotenuse), and therefore

$$A(\theta) = (6 + 6 \cos \theta)6 \sin \theta,$$

$$= 36(1 + \cos \theta)\sin \theta.$$

EXERCISES

Compute sin A, cos A, tan A, csc A, sec A, and cot A for the angle A defined in each of the following triangles.

1.

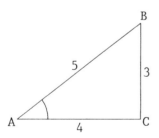

2.

3. Use a 45°-triangle to compute the values of the six trigonometric functions for A = 45°.

4. Use a 30°-60°-90° triangle to compute the values of the six trigonometric functions for A = 30° and A = 60°.

5. Evaluate $2 \sin x \cos x + \sin x$ for $x = \pi/4$ (radians).

6. Solve for y': $(\sec \frac{\pi}{3} \tan \frac{\pi}{3})y' = 2$.

141

7. Find the lengths of the two legs of the right triangle in the figure below.

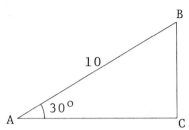

8. Find the lengths of the other two sides of the right triangle below (you will need a calculator).

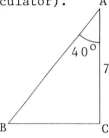

9. Consider the isosceles triangle ABC in the figure at the right. The equal sides AB and AC are of unit length, D is the foot of the altitude drawn from A, and E is the foot of the altitude drawn from C. Let θ denote the angle DAC, and assume that $0 < \theta < \pi/4$.

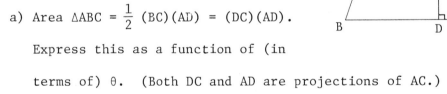

a) Area $\triangle ABC = \frac{1}{2}$ (BC)(AD) = (DC)(AD). Express this as a function of (in terms of) θ. (Both DC and AD are projections of AC.)

b) Area $\triangle ABC = \frac{1}{2}$ (AB)(EC) = $\frac{1}{2}$ (EC). Express this as a function of θ. (EC is a projection of AC.)

c) From a) and b), show that for $0 \leq \theta \leq \pi/4$, $\sin 2\theta = 2 \sin \theta \cos \theta$.

10. Consider again the isosceles triangle ABC described in Exercise 9. Let F be the midpoint of EB, and draw lines DF and DE. DF is parallel to EC (the

142

line joins the midpoints of adjacent sides of ΔBCE) and is perpendicular to AB. You can show that ΔBDF ~ ΔBAD and ΔBDF ≅ ΔEDF, and from this you can show that ED = BD and angle FDE = θ.

a) Express AE as a function of θ.

b) Express AF as a function of θ.

(AD is a projection of AC, and

AF is a projection of AD.)

c) Express EF as a function of θ.

(BD is a projection of AB and

EF is a projection of ED.)

d) Use the results of a), b), and c) to show that for $0 \leq \theta \leq \pi/4$,

$\cos 2\theta = \cos^2\theta - \sin^2\theta$.

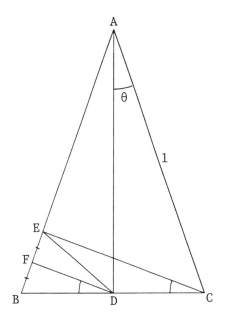

ANSWERS

	sin A	cos A	tan A	csc A	sec A	cot A
1.	3/5	4/5	3/4	5/3	5/4	4/3
2.	5/13	12/13	5/12	13/5	13/12	12/5
3.	$1/\sqrt{2}$	$1/\sqrt{2}$	1	$\sqrt{2}$	$\sqrt{2}$	1
4a.	1/2	$\sqrt{3}/2$	$1/\sqrt{3}$	2	$2/\sqrt{3}$	$\sqrt{3}$
4b.	$\sqrt{3}/2$	1/2	$\sqrt{3}$	$2/\sqrt{3}$	2	$1/\sqrt{3}$

5. $1 + \sqrt{2}/2$ 6. $1/\sqrt{3}$ 7. BC = 5, AC = $5\sqrt{3}$

8. BC = 7 tan 40° ≅ 5.8737,   AB = 7 sec 40° ≅ 9.13785

9. a) Area ΔABC = $\sin\theta\cos\theta$,   b) Area ΔABC = $\frac{1}{2}\sin 2\theta$

10. a) AE = $\cos 2\theta$, b) AF = $\cos^2\theta$ [= $(\cos\theta)^2$],   c) EF = $\sin^2\theta$

143

31. Trigonometric Functions of a General Angle

The six trigonometric functions defined in the previous section were defined only for angles between 0° and 90° (0 and $\pi/2$ radians). In this section
we extend their definitions so that they are defined for any arbitrary angle θ.
An angle is a measure of the amount of directed rotation (in degrees or radians)
from one half-line (initial side) to another half-line (terminal side). Any
angle θ can be put into a standard position in which the initial side corresponds with the positive x-axis, and the vertex of the angle lies at the origin
(see figure).

(The angle shown in the figure is roughly (360°) + (90° + 45°) = 495° or

(2π) + (3π/4) = 11π/4 radians.)

Let θ denote an arbitrary angle. Place the angle in standard position, and
choose any point P = (x,y) on the terminal side (different from the origin).
Let r denote the (positive) distance of P from the origin 0.

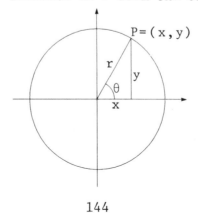

Define

$$\sin \theta = \frac{y}{r}, \qquad \csc \theta = \frac{r}{y},$$

$$\cos \theta = \frac{x}{r}, \qquad \sec \theta = \frac{r}{x},$$

$$\tan \theta = \frac{y}{x}, \qquad \cot \theta = \frac{x}{y}.$$

It is understood that the functions are left undefined if the denominator in their definition is zero. For example, if $\theta = \pi/2$, $x = 0$ so $\tan \pi/2$ and $\sec \pi/2$ are undefined; also when $\theta = 0$, $y = 0$ so $\csc 0$ and $\cot 0$ are undefined.

Observe that these definitions correspond to those given in the previous section if $0 < \theta < 90°$.

Example 1. Evaluate a) $\cos \frac{3\pi}{4}$, b) $\sin \frac{7\pi}{6}$, c) $\tan \frac{5\pi}{3}$, d) $\sec \frac{-\pi}{4}$.

Solution. These can be found by placing 45° or 30°-60°-90° triangles in appropriate quadrants. In each case begin by sketching the angle in standard position. For (a), the angle is as shown.

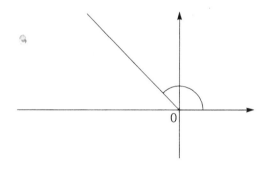

Choose $P = (-1,1)$ on the terminal side and form $\triangle OQP$ (drop a perpendicular from P to the x-axis to find Q).

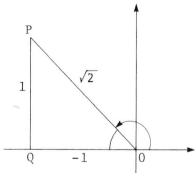

145

From the definition, $\cos \dfrac{3\pi}{4} = \dfrac{-1}{\sqrt{2}} = -\dfrac{\sqrt{2}}{2}$.

For $7\pi/6$ radians,

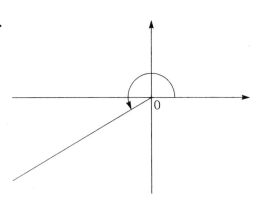

draw the 30°-60°-90° triangle as shown below.

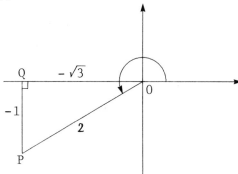

From the definition, $\sin \dfrac{7\pi}{6} = -\dfrac{1}{2}$.

To compute $\tan \dfrac{5\pi}{3}$, draw the 30°-60°-90° triangle as in the figure below.

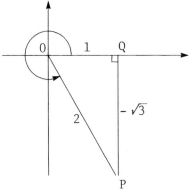

From the definition, $\tan \dfrac{5\pi}{3} = -\dfrac{\sqrt{3}}{1} = -\sqrt{3}$.

For $\sec (-\dfrac{\pi}{4})$, sketch the 45° triangle as shown from the definition $\sec (-\dfrac{\pi}{4}) = \dfrac{\sqrt{2}}{1} = \sqrt{2}$.

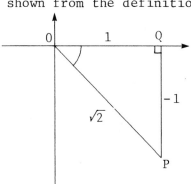

Example 2. Evaluate (a) sin 0 and cos 0; (b) sin π/2 and cos π/2;

(c) sin π and cos π.

Solution. For the case of an angle of 0 degrees (or radians), the terminal

side is the same as the initial side. Choose P = (1,0) on the terminal side.

Then, by definition, sin 0 = 0 and cos 0 = 1/1 = 1. For an angle of π/2 radi-

ans, choose P = (0,1) on the terminal side. Then sin π/2 = 1/1 = 1 and

cos π/2 = 0/1 = 0. For an angle of π radians, choose P = (-1,0) on the termi-

nal side. Then sin π = 0/1 = 0 and cos π = -1/1 = -1.

There is another way of thinking about the sine and the cosine which is

very insightful and helpful in computations such as those in Example 2. Let θ

be an arbitrary angle placed in standard position, and suppose that the termi-

nal side of this angle intersects the unit circle at the point P = (x,y). Then

the y-coordinate of P, namely y, is equal to the sine of θ (sin θ = $\frac{y}{1}$ = y), and

the x-coordinate of P, namely x, is the cosine of θ (cos θ = $\frac{x}{1}$ = x).

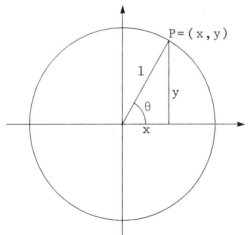

In other words, as P moves along the circumference, the sine and the cosine

of the corresponding angle θ are simply the vertical and horizontal projections

of the arm OP respectively. Thus, in Example 2, when P is moved down to (1,0),

the vertical projection of OP is 0, so that sin 0 = 0, and the horizontal

147

projection of OP is 1, so that cos 0 = 1. Similarly, for $\frac{\pi}{2}$ and π, compute the sine and cosine by considering the vertical and horizontal projections of OP, when P is moved to (0,1) and (-1,0) respectively. The next example gives another application of this important interpretation of the sine and cosine.

Example 3. a) Compare the values of: (i) sin θ and sin(-θ), (ii) cos θ and cos(π - θ).

b) Simplify (write as sin θ, -sin θ, cos θ, or - cos θ): (i) cos($\frac{\pi}{2}$ + θ), (ii) sin($\frac{\pi}{2}$ - θ).

Solution. a) The vertical projections of OP and OP' (see figure) are equal in magnitude but opposite in sign. It follows that sin(-θ) = - sin θ.

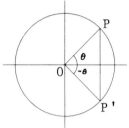

The horizontal projections of OP and OP' (see figure) are equal in magnitude but opposite in sign; therefore, cos(π - θ) = - cos θ.

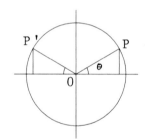

b) The horizontal projection of OP' (see figure) is equal in magnitude to the vertical projection of OP (congruent triangles) but opposite in sign. Thus, cos($\frac{\pi}{2}$ + θ) = - sin θ.

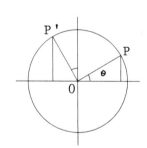

The vertical projection of OP' (see figure) is equal to the horizontal projection of OP; therefore, sin($\frac{\pi}{2}$ - θ) = cos θ.

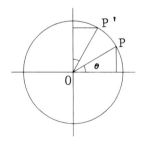

EXERCISES

Evaluate each of the following.

1. $\sin 5\pi/4$, $\cos 5\pi/4$, $\tan 5\pi/4$

2. $\sin 11\pi/6$, $\cos 11\pi/6$, $\tan 11\pi/6$

3. $\csc 4\pi/3$, $\sec 4\pi/3$, $\cot 4\pi/3$

4. $F(3\pi/4) - F(\pi/6)$, where $F(x) = \sin x$

5. Compare the values of: a) $\cos \theta$ and $\cos(-\theta)$, b) $\sin(\theta)$ and $\sin(\pi + \theta)$,

   c) $\sin \theta$ and $\cos(\theta - \frac{\pi}{2})$.

6. Suppose that $\frac{\pi}{4} \le \theta \le \frac{\pi}{2}$. Then $0 \le \frac{\pi}{2} - \theta \le \frac{\pi}{4}$, so by Exercises 9 and 10, Section 30, page 142,

$$\sin 2(\frac{\pi}{2} - \theta) = 2\sin(\frac{\pi}{2} - \theta)\cos(\frac{\pi}{2} - \theta),$$

$$\cos 2(\frac{\pi}{2} - \theta) = \cos^2(\frac{\pi}{2} - \theta) - \sin^2(\frac{\pi}{2} - \theta).$$

Show that these equations are equivalent to

$$\sin 2\theta = 2 \sin \theta \cos \theta,$$
$$\cos 2\theta = \cos^2\theta - \sin^2\theta.$$

7. $\sin 2\pi$, $\cos 2\pi$, $\tan 2\pi$

8. $\sin 3\pi/2$, $\cos 3\pi/2$

9. $2 \cos(\pi/2) - (\pi/2) \sin(\pi/2)$

10. $F(\pi/2) - F(0)$, where $F(x) = -x \sin x + \cos x$

ANSWERS

1. $\sin 5\pi/4 = -1/\sqrt{2}$, $\cos 5\pi/4 = -1/\sqrt{2}$, $\tan 5\pi/4 = 1$

2. $\sin 11\pi/6 = -1/2$, $\cos 11\pi/6 = \sqrt{3}/2$, $\tan 11\pi/6 = -1/\sqrt{3}$

3. $\csc 4\pi/3 = -2/\sqrt{3}$, $\sec 4\pi/3 = -2$, $\cot 4\pi/3 = 1/\sqrt{3}$

4. $1/\sqrt{2} - 1/2 = (\sqrt{2} - 1)/2$

5. a) $\cos(-\theta) = \cos\theta$,   b) $\sin(\pi + \theta) = -\sin\theta$,   c) $\cos(\theta - \pi/2) =$

$\cos(\pi/2 - \theta) = \sin\theta$ 6. $\sin 2(\pi/2 - \theta) = \sin(\pi - 2\theta) = \sin 2\theta$,

$\sin(\pi/2 - \theta) = \cos\theta$, $\cos(\pi/2 - \theta) = \sin\theta$, $\cos 2(\pi/2 - \theta) = \cos(\pi - 2\theta) =$

$-\cos 2\theta$.   7. $\sin 2\pi = 0$, $\cos 2\pi = 1$, $\tan 2\pi = 0$ 8. $\sin 3\pi/2 = -1$,

$\cos 3\pi/2 = 0$ 9. $-\pi/2$ 10. $-\pi/2 - 1$

---

32.   Graphs of Sine and Cosine

The graph of $f(\theta) = \sin\theta$ can be sketched in the usual way by constructing a table and carefully plotting the points.

θ	0	$\pi/6$	$\pi/4$	$\pi/3$	$\pi/2$	$2\pi/3$	etc.
$f(\theta)$	0	1/2	$\sqrt{2}/2$	$\sqrt{3}/2$	1	$\sqrt{3}/2$	

The result is the graph below.

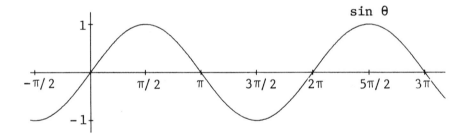

Another way to determine the shape of this graph is to consider the motion of the y-coordinate of a point $P = (x,y)$ moving counterclockwise around the unit circle. The y-coordinate of P is the sine of the angle θ (see following figure). As P moves along the circumference through the first quadrant, $\sin\theta$ goes from 0 to 1, and then decreases to 0 again as P moves through the second quadrant to the point $(-1,0)$. As P moves through the third and fourth

150

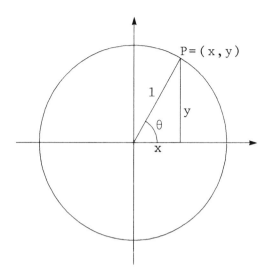

quadrants the y-coordinate is negative, and will reach a minimum value of -1 at $\theta = 3\pi/2$. Each time P retraces the circle, the cycle will be repeated; we say that sin θ is *periodic*, of period 2π; that is, for all θ,

$$\sin(\theta + 2\pi) = \sin \theta.$$

The graph of $f(\theta) = \cos \theta$ can be sketched in a similar way. The x-coordinate of P is the cosine of θ (see preceding figure). As P moves around the unit circle from (1,0) to (0,1), cos θ moves from 1 to 0; cos θ is negative as P circles through quadrants two and three, reaching a minimum value of -1 at $\theta = \pi$, then rises again to 1 as P moves from (0,-1) to (1,0). The graph is shown below.

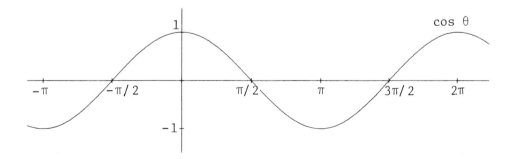

151

This graph is also periodic, of period 2π; that is, for all θ,

$$\cos(\theta + 2\pi) = \cos \theta.$$

Example 1. Sketch the graph of $f(\theta) = 3 \sin \theta$.

Solution. For a given θ, the y-coordinate of a point on $y = 3 \sin \theta$ is simply three times the y-coordinate, for the same θ, of a point on $y = \sin \theta$.

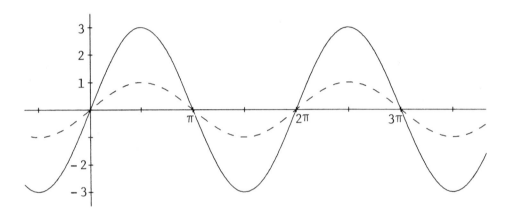

In a similar manner, the graph of $y = A \sin \theta$ (or, $y = A \cos \theta$) can be obtained from that of $y = \sin \theta$ (or, $y = \cos \theta$) (by magnifying or shrinking the y-coordinates of $y = \sin \theta$ by the factor A). The magnitude of A, $|A|$, is called the *amplitude*; it is a measure of the size of the oscillation (maximum value of the function).

Example 2. Sketch the graph of $f(\theta) = \cos(\theta/2)$.

Solution. We know that $\cos(\theta/2)$ will complete one period (one full cycle) when its argument, $\theta/2$, varies from 0 to 2π. That is, we have one period when

$$0 \leq \theta/2 \leq 2\pi, \quad \text{or equivalently,} \quad 0 \leq \theta \leq 4\pi.$$

152

Thus, cos(θ/2) completes one period as θ varies from 0 to 4π. Therefore the graph is as shown below.

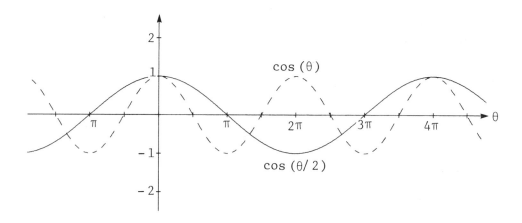

Example 3.   Sketch the graph of f(θ) = 2 cos(2θ + π/2).

Solution.   The graph completes one cosine cycle when

$$0 \leq 2\theta + \pi/2 \leq 2\pi.$$

Subtracting π/2 and dividing by 2, we find that the graph goes through one period when

$$-\pi/4 \leq \theta \leq \frac{2\pi - \pi/2}{2} = 3\pi/4.$$

This inequality makes it easy to sketch the graph shown below.

[2 cos(2θ + π/2) has amplitude 2, period π, and is shifted π/4 to the left.]

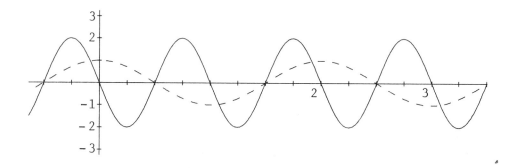

EXERCISES

Sketch graphs for the following trigonometric functions. In each case, start with the basic graph $y = \sin \theta$ or $y = \cos \theta$ and adjust the period and amplitude as in the previous examples.

1. $f(\theta) = 2 \cos \theta$
2. $f(\theta) = - \sin \theta$
3. $f(\theta) = \sin 2\theta$
4. $f(\theta) = 3 \sin(\theta/2)$
5. $f(\theta) = 2 \cos 3\theta$
6. $f(\theta) = \sin(\theta + \pi)$
7. $f(\theta) = 2 \cos(\theta - \pi/2)$
8. $f(\theta) = \frac{1}{2} \sin(\theta - \pi)$
9. $f(\theta) = 2 \sin(2\theta + \pi)$
10. $f(\theta) = 3 \cos(\theta/2 - \pi/6)$

ANSWERS

1.

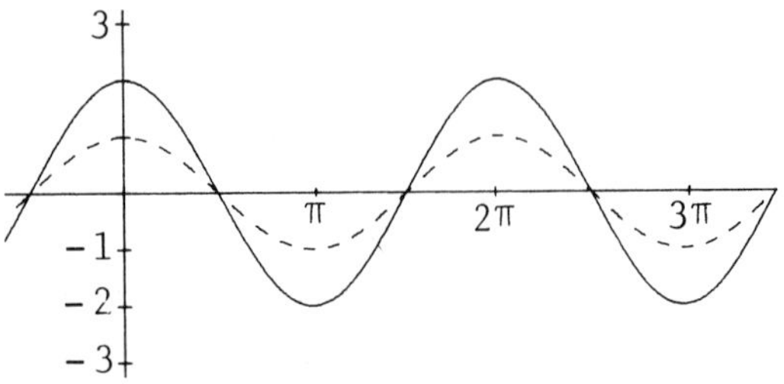

2.

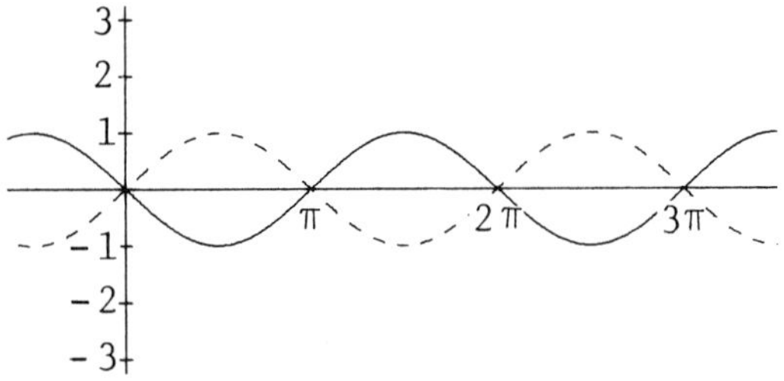

3.

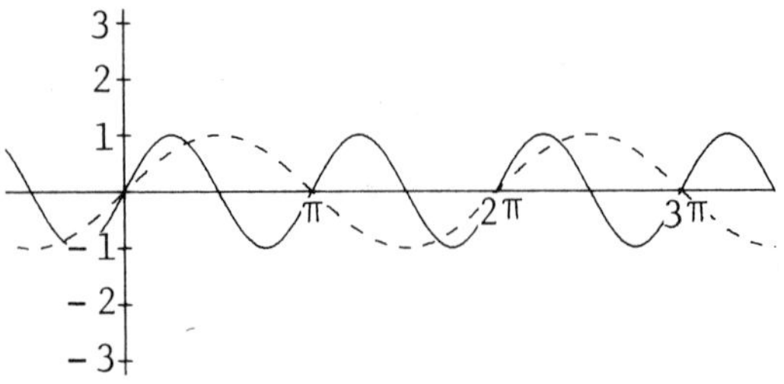

4.

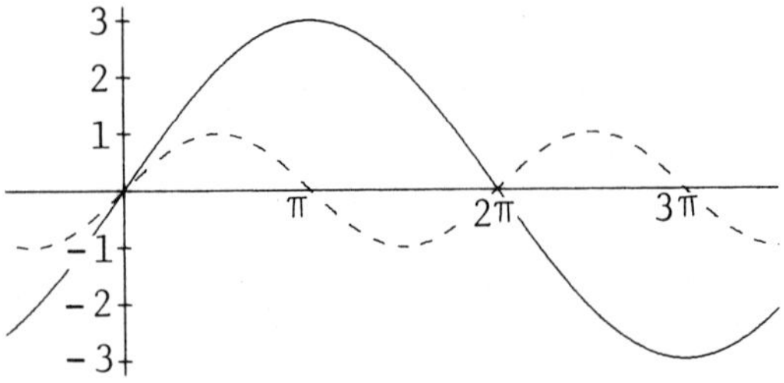

5.

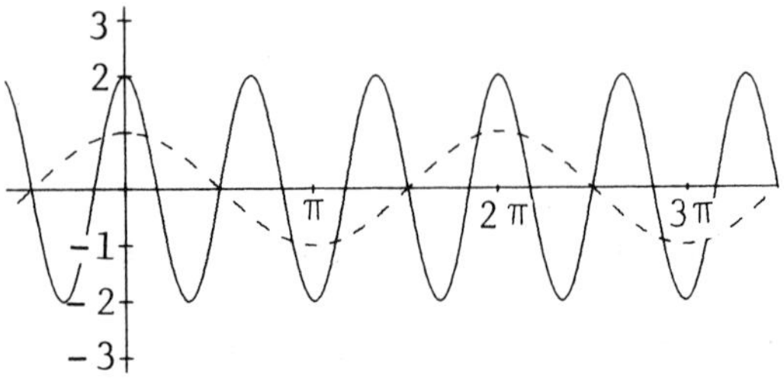

6.

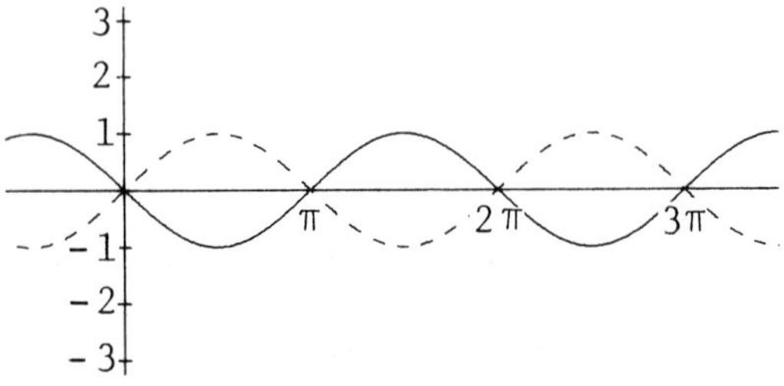

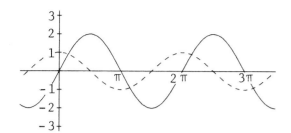

8.

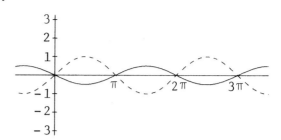

9.

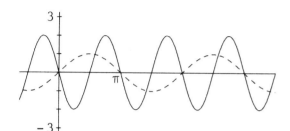

10.

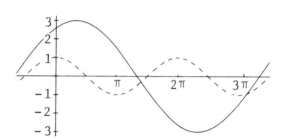

---

33. Trigonometric Equations

An equation containing trigonometric functions is called a trigonometric equation; solving such an equation consists of finding the angle (usually *angles*) which satisfies the equation.

Example 1. Solve for θ in each of the following trigonometric equations.

   a) $\sin \theta = 0$

   b) $\cos 2\theta = -1/2$

Solution. As θ varies from 0 to 2π, $\sin \theta$ will take on the value 0 twice, at $\theta = 0$ and $\theta = \pi$. Because sine has period 2π, the solutions to a) are

$\theta = n\pi, \quad n = 0, \pm 1, \pm 2, \ldots.$

As 2θ varies from 0 to 2π, $\cos 2\theta$ will take on the value $-1/2$ when $2\theta = 2\pi/3$ and $2\theta = 4\pi/3$. Because cosine has period 2π, we will also have solutions at $2\theta = 2\pi/3 + n \cdot 2\pi$ and $2\theta = 4\pi/3 + n \cdot 2\pi$ for $n = 0, \pm 1, \pm 2, \ldots$. Solving for θ, we find the solutions to b) are

$$\theta = \frac{1}{3}\pi + n\pi, \qquad n = 0, \pm 1, \pm 2, \ldots, \quad \text{and}$$

$$\theta = \frac{2}{3}\pi + n\pi, \qquad n = 0, \pm 1, \pm 2, \ldots.$$

Example 2. Find all solutions to $2\cos^2\theta + \cos\theta - 1 = 0$.

Solution. This is a quadratic equation in $\cos\theta$. In this case, the quadratic can be factored as $(2\cos\theta - 1)(\cos\theta + 1) = 0$. The solutions can therefore be broken down into two types: those θ for which $\cos\theta = -1$ and those θ for which $\cos\theta = 1/2$. Arguing as in Example 1, we find that that $\cos\theta = -1$ when

$$\theta = \pi + 2n\pi, \qquad n = 0, \pm 1, \pm 2, \ldots,$$

and $\cos\theta = 1/2$ when

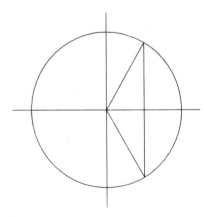

$$\theta = \frac{\pi}{3} + 2n\pi, \qquad n = 0, \pm 1, \pm 2, \ldots, \quad \text{and}$$

$$\theta = -\frac{\pi}{3} + 2n\pi, \qquad n = 0, \pm 1, \pm 2, \ldots.$$

Example 3. Find all solutions to $\sin\theta = \cos\theta$.

Solution. We may presume that $\cos\theta \neq 0$, since no angle which makes $\cos\theta = 0$ will also make $\sin\theta = 0$. Therefore, we can write the equation in the form

$$\frac{\sin\theta}{\cos\theta} = 1, \quad \text{or equivalently,} \quad \tan\theta = 1.$$

$$\left(\frac{\sin \theta}{\cos \theta} = \frac{y/r}{x/r} = \frac{y}{x} = \tan \theta. \right)$$

Geometrically, tan θ is equal to the *slope* of the terminal side of the angle (this is because tan θ = y/x = slope of OP).

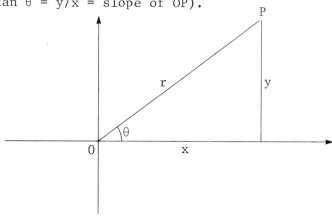

As θ varies from −π/2 to π/2, the slope of line OP equals 1 when, and only when, θ = π/4. Thus, the only solution of tan θ = 1 in the interval from −π/2 to π/2 is θ = π/4.

For other solutions, note that tan θ is periodic, of period π: for all θ,

$$\tan(\theta + \pi) = \tan \theta.$$

The graph of f(θ) = tan θ is as shown below.

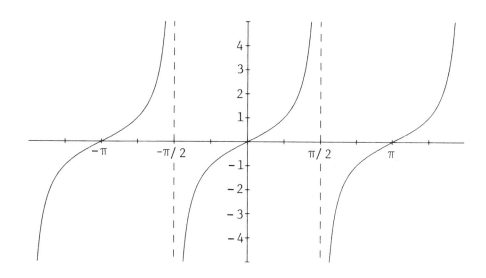

157

Find all solutions to the following trigonometric equations.

1. $\cos \theta = 0$

2. $\sin \theta = -1$

3. $\cos \theta = \sqrt{2}/2$

4. $\cot \theta = -\sqrt{3}$

5. $\tan 2\theta = -1$

6. $\sin \theta \cos \theta + \sin \theta = 0$

7. $\cos^2\theta - 2 \cos \theta + 1 = 0$

8. $2 \sin^2\theta - \sin \theta - 1 = 0$

9. $\sin^2\theta = 1/4$

10. $\sqrt{3} \cos \theta = \sin \theta$

ANSWERS

1. $\theta = \pi/2 + n\pi, \ n = 0, \pm 1, \pm 2, \ldots$

2. $\theta = 3\pi/2 + 2n\pi, \ n = 0, \pm 1, \pm 2, \ldots$

3. $\theta = \pi/4 + 2n\pi, \ n = 0, \pm 1, \pm 2, \ldots,$ and $\theta = 7\pi/4 + 2n\pi, \ n = 0, \pm 1, \pm 2, \ldots$

4. $\theta = 5\pi/6 + n\pi, \ n = 0, \pm 1, \pm 2, \ldots$

5. $\theta = -\pi/8 + n\pi/2, \ n = 0, \pm 1, \pm 2, \ldots$

6. $\theta = n\pi, \ n = 0, \pm 1, \pm 2, \ldots$

7. $\theta = 2n\pi, \ n = 0, \pm 1, \pm 2, \ldots$

8. $\theta = \pi/2 + 2n\pi, \ n = 0, \pm 1, \pm 2, \ldots,$ and $\theta = -\pi/6 + 2n\pi, \ n = 0, \pm 1, \pm 2, \ldots,$ and $\theta = 7\pi/6 + 2n\pi, \ n = 0, \pm 1, \pm 2, \ldots$

9. $\theta = \pm\pi/6 + n\pi, \ n = 0, \pm 1, \pm 2, \ldots$

10. $\theta = \pi/3 + n\pi, \ n = 0, \pm 1, \pm 2, \ldots$

---

34. Trigonometric Identities

There are a number of trigonometric identities that enable you to simplify complicated trigonometric expressions. To make this list complete, we will include five formulas in this list that have been introduced in previous sections:

$$\csc \theta = \frac{1}{\sin \theta} , \qquad (1)$$

$$\sec \theta = \frac{1}{\cos \theta} , \qquad\qquad (2)$$

$$\cot \theta = \frac{1}{\tan \theta} , \qquad\qquad (3)$$

$$\tan \theta = \frac{\sin \theta}{\cos \theta} , \qquad\qquad (4)$$

$$\cot \theta = \frac{\cos \theta}{\sin \theta} . \qquad\qquad (5)$$

(These identities and those that follow are valid for all values of the angle for which each side of the identity is defined.) In the list that follows, the starred (*) formulas are the most basic in the sense that the others can be derived directly from them, so you might focus on learning them first.

The next set of three formulas are called the Pythagorean identites.

$$(*) \qquad\qquad \sin^2\theta + \cos^2\theta = 1 , \qquad\qquad (6)$$

$$\tan^2\theta + 1 = \sec^2\theta, \qquad\qquad (7)$$

$$1 + \cot^2\theta = \csc^2\theta. \qquad\qquad (8)$$

The reason for this identification is that they are each equivalent to the Pythagorean Theorem; for let θ be an arbitrary angle placed in standard position, and let the terminal side of θ intersect the unit circle in the point $P = (x,y)$.

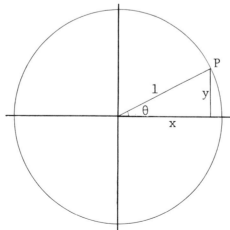

By the Pythagorean Theorem, $x^2 + y^2 = 1$. But $x = \cos\theta$ and $y = \sin\theta$ and

therefore $x^2 + y^2 = 1$ is equivalent to (6). When we divide each side of (6) by

$\cos^2\theta$, we get

$$\frac{\sin^2\theta}{\cos^2\theta} + \frac{\cos^2\theta}{\cos^2\theta} = \frac{1}{\cos^2\theta} \, ,$$

and this is the same as (7). Similarly, if we divide (6) by $\sin^2\theta$ and simpli-

fy, we get (8).

These eight formulas are often called the *eight fundamental identities of*

trigonometry. They allow us to express any trigonometric expression in terms of

a single trigonometric function (see Example 1), and this is useful in solving

trigonometric equations.

Example 1. Express $\dfrac{\cot\theta - \sin\theta\cos\theta}{1 + 2\csc\theta}$ purely in terms of $\sin\theta$.

Solution. A common simplification technique used in working with trigonometric

functions is to reduce the expression to one involving only sines and cosines.

In this case, the given expression is equivalent to

$$\frac{\dfrac{\cos\theta}{\sin\theta} - \sin\theta\cos\theta}{1 + \dfrac{2}{\sin\theta}} \, ,$$

$$\frac{\cos\theta - \sin^2\theta\cos\theta}{\sin\theta} \cdot \frac{\sin\theta}{\sin\theta + 2} \, ,$$

$$\frac{\cos\theta\,(1 - \sin^2\theta)}{2 + \sin\theta} \, .$$

Using (6), we may write this as

$$\frac{\cos^3\theta}{2 + \sin\theta} \, .$$

Again, by (6), $\cos\theta = \pm(1 - \sin^2\theta)^{1/2}$; the choice of sign depends on the value

of θ. If $-\pi/2 \leq \theta \leq \pi/2$, then $\cos\theta$ is positive, and we take the positive

160

square root:

$$\frac{(1 - \sin^2\theta)^{3/2}}{2 + \sin\theta}\ .$$

If $\pi/2 < \theta < 3\pi/2$, $\cos\theta$ is negative so we take the negative square root and find that the original expression is the negative of the last expression.

The following *addition formulas* are basic:

(*) $\sin(\alpha \pm \beta) = \sin\alpha\,\cos\beta \pm \cos\alpha\,\sin\beta,$ (9)

(*) $\cos(\alpha \pm \beta) = \cos\alpha\,\cos\beta \mp \sin\alpha\,\sin\beta,$ (10)

$$\tan(\alpha \pm \beta) = \frac{\tan\alpha \pm \tan\beta}{1 \mp \tan\alpha\,\tan\beta}\ .$$ (11)

The proofs of these identities are rather technical and will not concern us here.

Example 2. Evaluate $\sin 75°$ and $\cos 75°$.

Solution. Set $\alpha = 45°$ and $\beta = 30°$. From the addition formulas,

$$\sin 75° = \sin(45° + 30°) = \sin 45°\,\cos 30° + \cos 45°\,\sin 30°$$

$$= \left(\frac{\sqrt{2}}{2}\right)\left(\frac{\sqrt{3}}{2}\right) + \left(\frac{\sqrt{2}}{2}\right)\left(\frac{1}{2}\right)$$

$$= \frac{\sqrt{2}\,(\sqrt{3} + 1)}{4}\ ,$$

and

$$\cos 75° = \cos(45° + 30°) = \cos 45°\,\cos 30° - \sin 45°\,\sin 30°$$

$$= \left(\frac{\sqrt{2}}{2}\right)\left(\frac{\sqrt{3}}{2}\right) - \left(\frac{\sqrt{2}}{2}\right)\left(\frac{1}{2}\right)$$

$$= \frac{\sqrt{2}\,(\sqrt{3} - 1)}{4}\ .$$

By setting $\alpha = \theta$ and $\beta = \theta$ in equations (9) and (10), we get the important *double angle formulas:*

$$\sin 2\theta = 2 \sin \theta \cos \theta, \qquad\qquad (12)$$

$$\cos 2\theta = \cos^2 \theta - \sin^2 \theta. \qquad\qquad (13)$$

(Proofs of these identities for $0 \le \theta \le \pi/2$ were outlined in Exercises 9 and 10,

Section 30, and Exercise 6, Section 31.)

When we add equations (6) and (13), we get

$$(\sin^2 \theta + \cos^2 \theta) + (\cos^2 \theta - \sin^2 \theta) = 1 + \cos 2\theta,$$

$$2 \cos^2 \theta = 1 + \cos 2\theta,$$

$$\cos^2 \theta = \frac{1 + \cos 2\theta}{2}.$$

If we subtract equation (13) from equation (6) we get

$$(\sin^2 \theta + \cos^2 \theta) - (\cos^2 \theta - \sin^2 \theta) = 1 - \cos 2\theta,$$

$$2 \sin^2 \theta = 1 - \cos 2\theta,$$

$$\sin^2 \theta = \frac{1 - \cos 2\theta}{2}.$$

In these two equations for $\sin^2 \theta$ and $\cos^2 \theta$, we replace θ by $\theta/2$ to get the

half angle formulas:

$$\sin^2 \frac{\theta}{2} = \frac{1 - \cos \theta}{2}, \qquad\qquad (14)$$

$$\cos^2 \frac{\theta}{2} = \frac{1 + \cos \theta}{2}. \qquad\qquad (15)$$

The double angle formulas and the half angle formula are useful in calculus

when you study methods of integration.

To this core list of fifteen trigonometric identities we add two others:

the *law of sines* and the *law of cosines*. These state that in an *arbitrary*

triangle ABC:

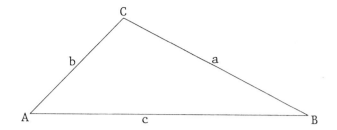

(Law of Sines)   (*) $\dfrac{\sin A}{a} = \dfrac{\sin B}{b} = \dfrac{\sin C}{c}$ (16)

(Law of Cosines)   (*) $a^2 = b^2 + c^2 - 2bc \cos A$ (17)

$$b^2 = a^2 + c^2 - 2ac \cos B$$

$$c^2 = a^2 + b^2 - 2ab \cos C$$

The law of cosines is a generalization of the Pythagorean Theorem, for when C is 90°, we get

$$c^2 = a^2 + b^2 - 2ab \cos 90° = a^2 + b^2.$$

Example 3.   Two sides of a triangle are 1 and 4 respectively and the angle between them is 60°. Find the length of the third side, and the measures of the other angles of the triangle.

Solution.   To find the length of the third side, we apply the law of cosines:

$c^2 = 1^2 + 4^2 - 2 \cdot 1 \cdot 4 \cos(60°)$

     $= 17 - 8(\frac{1}{2})$

     $= 13,$

so that $c = \sqrt{13}.$

Using the law of sines,

$$\frac{\sin B}{1} = \frac{\sin C}{c}$$

$$\sin B = \frac{\sqrt{3}/2}{\sqrt{13}} = \frac{1}{2}\sqrt{\frac{3}{13}} \approx .24019223.$$

The angle whose sine is .24019223 is 13.897886° (see Section 35). Therefore,

$$A = 180° - (60° + 13.897887°) = 106.10211°.$$

EXERCISES

Prove the following identities.

1. $\tan \theta + \sec \theta = \dfrac{1 + \sin \theta}{\cos \theta}$

2. $\dfrac{\cot \theta}{\csc \theta - 1} = \dfrac{1 + \sin \theta}{\cos \theta}$

3. $\dfrac{1 + \tan^2 \theta}{\csc \theta} = \sec \theta \tan \theta$

4. Show that $\dfrac{\sin(x + h) - \sin x}{h} = \left(\dfrac{\cos h - 1}{h} \right) \sin x + \left(\dfrac{\sin h}{h} \right) \cos x.$

5. Prove one of the following identities which express products of sines and cosines as sums.

   a) $\sin \alpha \cos \beta = \dfrac{1}{2} [\sin(\alpha + \beta) + \sin(\alpha - \beta)]$

   b) $\cos \alpha \sin \beta = \dfrac{1}{2} [\sin(\alpha + \beta) - \sin(\alpha - \beta)]$

   c) $\cos \alpha \cos \beta = \dfrac{1}{2} [\cos(\alpha + \beta) + \cos(\alpha - \beta)]$

   d) $\sin \alpha \sin \beta = -\dfrac{1}{2} [\cos(\alpha + \beta) - \cos(\alpha - \beta)]$

6. If $\cos \theta = \dfrac{24}{25}$ and $0 < \theta < \pi/2$, evaluate $\sin 2\theta$ and $\cos 2\theta$.

7. If $\sin \theta = \dfrac{3}{5}$ and $0 < \theta < \pi/2$, find $\sin \dfrac{\theta}{2}$ and $\cos \dfrac{\theta}{2}$.

8. Solve the equation $\cos 2\theta - \sin \theta = 0$, by first changing the equation to a quadratic equation in $\sin \theta$.

9. Use the law of sines to find the length of the other sides of the triangle below.

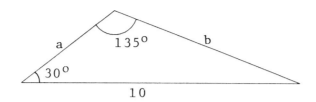

10. The angle subtended by the boundaries of the shot put and discus throwing sector is 40°. If you have two tape measures, how would you mark out these boundaries?

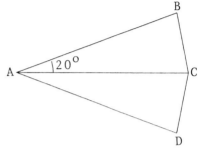

b) To make a rough check on the angle, suppose B and D are 25 paces from A. For an angle of 40°, how many paces should there be between B and D?

ANSWERS

1. $\tan\theta + \sec\theta = \dfrac{\sin\theta}{\cos\theta} + \dfrac{1}{\cos\theta} = \dfrac{1 + \sin\theta}{\cos\theta}$

2. $\dfrac{\cot\theta}{\csc\theta - 1} = \dfrac{\cos\theta/\sin\theta}{(1/\sin\theta) - 1} = \dfrac{\cos\theta}{1 - \sin\theta} \cdot \dfrac{1 + \sin\theta}{1 + \sin\theta} = \dfrac{\cos\theta(1 + \sin\theta)}{\cos^2\theta}$

$= \dfrac{1 + \sin\theta}{\cos\theta}$ 3. $\dfrac{1 + \tan^2\theta}{\csc\theta} = \dfrac{\sec^2\theta}{\csc\theta} = \dfrac{\sin\theta}{\cos^2\theta} = \dfrac{\sin\theta}{\cos\theta} \cdot \dfrac{1}{\cos\theta} =$

$\tan\theta \sec\theta$ 4. Use (9), expand and simplify.   5. Expand right hand side using equations (9) and (10), then simplify.

6. $\sin 2\theta = 336/625$, $\cos 2\theta = 527/625$ 7. $\sin\dfrac{\theta}{2} = \dfrac{\sqrt{10}}{10}$, $\cos\dfrac{\theta}{2} = \dfrac{3\sqrt{10}}{10}$

8. $\theta = 3\pi/2 + 2n\pi$, $n = 0,\pm1,\pm2,\ldots$, and $\theta = \pi/6 + 2n\pi$, $n = 0,\pm1,\pm2,\ldots$, and

$\theta = 5\pi/6 + 2n\pi$, $n = 0,\pm1,\pm2,\ldots$ 9. $b = 5\sqrt{2}$, $a = 5\sqrt{2(2 - \sqrt{3})}$

10. One approach is to choose B so that AB = AC and BC is such that ∠CAB = 20°

(see figure). By the law of cosines, x is

given by the equation

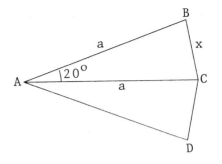

$$x^2 = a^2 + a^2 - 2a^2\cos 20°$$

$$= 2a^2[1 - \cos 20°]$$

so that $x = a\sqrt{2(1 - \cos 20°)} = .34729635a$.

When a = 100 ft, x = 34.729635 $\cong$ 34 ft 8 3/4 in.   b) Using the law of cosines,

we find that BD = $25\sqrt{2(1 - \cos 40°)}$ $\cong$ 17.1 paces.

---

35. Inverse Trigonometric Functions

If x is any number between −1 and 1 inclusive, there is a unique angle θ

between $-\pi/2$ and $\pi/2$ (radians) such that sin θ = x. We denote this angle θ by

Arcsin x.   In other words, Arcsin x *is the angle (between $-\pi/2$ and $\pi/2$) whose*

sine is x.   ("Arc" translates into "the angle whose.")   If we set θ = Arcsin x,

then sin θ = x; the relation between θ and x is as in the figure below:

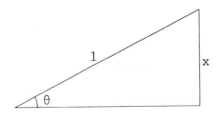

Example 1.   Evaluate   a) Arcsin($\sqrt{2}/2$),   b) Arcsin(−1/2).

Solution.   Arcsin($\sqrt{2}/2$) is the angle between $-\pi/2$ and $\pi/2$ whose sine is $\sqrt{2}/2$,

and this is $\pi/4$.

Arcsin(−1/2) is the angle between $-\pi/2$ and $\pi/2$ whose sine is −1/2, and

166

and this angle is $-\pi/6$ (see figure).

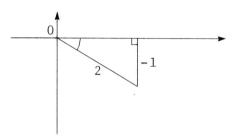

Arccos x and Arctan x are defined in a similar way: if x is between -1 and 1 inclusive, there is a unique angle θ between 0 and π (*not* $-\pi/2$ and $\pi/2$) such that $\cos \theta = x$; we denote this angle by Arccos x. If x is *any* real number, there is a unique angle θ between $-\pi/2$ and $\pi/2$ (again) such that $\tan \theta = x$; we denote this angle by Arctan x.

The three functions Arcsin, Arccos, and Arctan are *inverse trigonometric functions*. They are also denoted by $\sin^{-1}$, $\cos^{-1}$, and $\tan^{-1}$. Thus $\sin^{-1} 1$ means Arcsin 1 which equals $\pi/2$ (the angle whose sine is 1 is $\pi/2$) and *not* $\dfrac{1}{\sin 1}$.

Example 2. Evaluate

    a) $\tan(\text{Arccos}(-3/5)$

    b) $\sin(\text{Arccos } x)$

    c) $\cos(\text{Arcsin } x + \text{Arctan}(1/x))$, $0 < x < 1$.

Solution. It is always helpful when working with inverse trigonometric functions to draw a triangle showing the angles in question and their relationship to the sides. For a) we first sketch Arccos(-3/5). This angle is in the second quadrant, and the cosine of the angle is -3/5 (see figure).

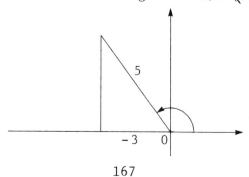

The third side of the relevant triangle is 4. The tangent of this angle is

4/(-3). Thus

$$\tan(\text{Arccos}(-3/5)) = -(4/3).$$

For b), let $\theta = \text{Arccos } x$.

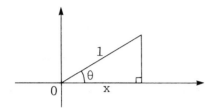

θ is the angle whose

cosine is x.

From the Pythagoream Theorem, the vertical side equals $\sqrt{1 - x^2}$ (the radical

is positive for all values of x, since $0 \leq \text{Arccos } x \leq \pi$). Thus

$$\sin(\text{Arccos } x) = \sqrt{1 - x^2}.$$

For c) let $\alpha = \text{Arcsin } x$ and $\beta = \text{Arctan}(1/x)$ (see figures below).

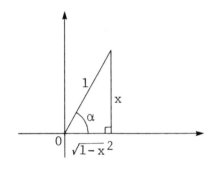

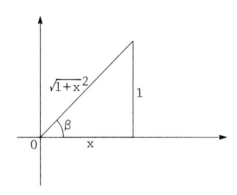

From the addition formula for cosine,

$$\cos(\alpha + \beta) = \cos \alpha \cos \beta - \sin \alpha \sin \beta.$$

Substituting into this identity we have

$$\cos(\alpha + \beta) = \sqrt{1 - x^2} \cdot \frac{x}{\sqrt{1 + x^2}} - x \frac{1}{\sqrt{1 + x^2}}$$

$$= \frac{x(\sqrt{1 - x^2} - 1)}{\sqrt{1 + x^2}}.$$

168

Example 3. In the figure below, express θ as a function of x. The angle θ is

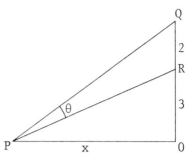

the angle subtended by the eye when viewing a 2-foot picture hanging three feet above eye level when standing x feet from the wall. Make a graph of the function above by plotting values of θ for x = 1,2,...,10 to determine the best viewing distance (the largest value of θ).

Solution. Let α = ∠OPQ and β = ∠OPR. Then

$$\theta = \alpha - \beta = \text{Arctan } \frac{5}{x} - \text{Arctan } \frac{3}{x}.$$

The table below can be constructed with the use of a hand calculator.

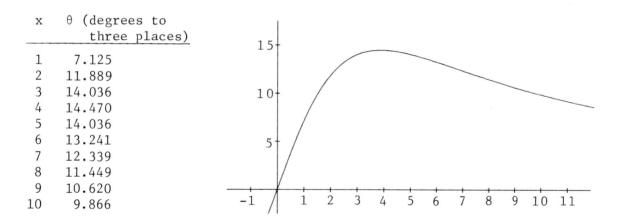

x	θ (degrees to three places)
1	7.125
2	11.889
3	14.036
4	14.470
5	14.036
6	13.241
7	12.339
8	11.449
9	10.620
10	9.866

These computations indicate that the optimal viewing distance is about 4 feet from the wall. (It may be of interest to mention here that the theory of calculus allows us to say that the largest value for θ will occur when x is such that $\frac{-5/x^2}{1 + (5/x)^2} - \frac{-3/x^2}{1 + (3/x)^2} = 0.$ The solutions of this equation are x = $\pm\sqrt{15}$ (see Exercise 8, Section 17). Thus, the optimal viewing distance occurs

$\sqrt{15}$ feet from the wall, at which point θ is approximately 14.478 degrees.

<u>Example 4</u>. Solve for x in the equation $y = 3 - 2 \sin(\frac{x+1}{2})$.

<u>Solution</u>. We begin by solving for $\sin(\frac{x+1}{2})$:

$$\sin(\frac{x+1}{2}) = \frac{3-y}{2}.$$

Taking the Arcsin of each side we get

$$\text{Arcsin}(\sin(\frac{x+1}{2})) = \text{Arcsin}(\frac{3-y}{2}).$$

But $\text{Arcsin}(\sin(\frac{x+1}{2})) = \frac{x+1}{2}$.   Therefore,

$$\frac{x+1}{2} = \text{Arcsin}(\frac{3-y}{2}),$$

$$x = -1 + 2\,\text{Arcsin}(\frac{3-y}{2}).$$

EXERCISES

Evaluate each of the following.

1. Arcsin 0

2. Arcsin(-1)

3. Arccos(-√3/2)

4. Arctan(-1)

5. cos(Arctan 2)

6. tan(Arcsin x)

7. cos(Arctan x)

8. cos(Arcsin 3/5 + Arctan 5/12)

9. sin(Arccos x + Arctan x)

10. One of the primary uses of algebra and trigonometry is to manipulate unwieldy expressions into more appropriate and understandable forms. For example: (i) What is the minimum value of $x^2 - 4x + 7$? Rewrite the expression as $(x-2)^2 + 3$, and it is clear that the minimum is 3. (ii) For what values of x, y, and z is xy + z = y + xz? Make the following changes in the form: (xy - y) + (z - xz) = 0; y(x - 1) + z(1 - x) = 0; (x - 1)(y - z) = 0. Now the answer is

clear: $x = 1$ or $y = z$. (iii) If $0 < \theta < \pi$, is

$$\text{Arctan}\left(\frac{1 - \cos \theta}{\sin \theta} \right) + \text{Arctan}\left(\frac{\cos \theta}{\sin \theta} \right)$$

more or less than $\pi/2$? The steps below lead to the answer.

a) Denote the expression by y. Show that $\tan y$ is equal to $\dfrac{\sin \theta}{1 - \cos \theta}$ (apply identity (11) of Section 34 and simplify).

b) Verify each of the following equalities:

$$\frac{\sin \theta}{1 - \cos \theta} = \frac{2 \sin \theta/2 \cos \theta/2}{2 \sin^2 \theta/2} = \cot \frac{\theta}{2} = \tan\left(\frac{\pi}{2} - \frac{\theta}{2} \right).$$

c) Use parts (a) and (b) to prove that $y = \pi/2 - \theta/2$.

ANSWERS

1. 0   2. $-\pi/2$ 3. $5\pi/6$ 4. $-\pi/4$ 5. $\sqrt{5}/5$

6. $\dfrac{x}{\sqrt{1 - x^2}}$ 7. $\dfrac{1}{\sqrt{1 + x^2}}$ 8. $33/65$ 9. $\dfrac{\sqrt{1 - x^2} + x^2}{\sqrt{1 + x^2}}$

36. Complex Numbers

Any number of the form $a + bi$, $i = \sqrt{-1}$, where a and b are real numbers, is called a *complex number*. Complex numbers arise naturally as solutions to quadratic equations (this is why they were invented in the first place). For example, the solutions to the quadratic equation

$$z^2 + z + 1 = 0$$

are (from the quadratic formula)

$$z = \frac{-1 \pm \sqrt{1 - 4}}{2}$$

171

$$= -\frac{1}{2} \pm \frac{\sqrt{3}}{2} i.$$

Complex numbers can be added, subtracted, multiplied, and divided according to the usual operations, except that whenever i^2 occurs, it is replaced by -1.

Example 1. Write each of the numbers below in the form $a + bi$, a and b real.

    a) $(3 + 2i) + (4 - i) - (5 - 2i)$

    b) $(4 + i)(3 - 4i)$

    c) $\dfrac{2 + 3i}{5 - 2i}$

Solution. For a) combine real parts and imaginary parts to get $2 + 3i$. For b), multiply in the usual way to get $12 + 3i - 16i - 4i^2$ which, when i^2 is replaced by -1, simplifies to $16 - 13i$. For c), multiply numerator and denominator by $5 + 2i$ (the conjugate of the denominator):

$$\left(\frac{2 + 3i}{5 - 2i}\right)\left(\frac{5 + 2i}{5 + 2i}\right) = \frac{10 + 19i - 6}{25 + 4}$$

$$= \frac{4}{29} + \frac{19}{29} i.$$

The complex number $a + bi$ can be represented geometrically in the rectangular coordinate system by the *vector* $\overrightarrow{OP}$, or arrow, from the origin 0 to the point $P = (a,b)$.

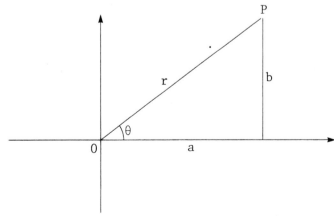

The *magnitude* (*modulus, absolute value*) of the complex number a + bi is defined to be the length of the vector $\overrightarrow{OP}$, and it is written as $|a + bi|$. Thus

$$|\overrightarrow{OP}| \equiv |a + bi| \equiv \sqrt{a^2 + b^2} \ .$$

The angle θ between the positive x-axis and the vector $\overrightarrow{OP}$ is called the *argument* of a + bi; it is determined only within a multiple of 2π.

The complex number z = $-\sqrt{3}$ + i is represented geometrically in the diagram below.

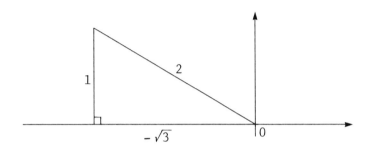

Its magnitude is 2 (= $\sqrt{(-\sqrt{3})^2 + (1)^2}$), and its argument is 5π/6 (+ 2πk).

Suppose that z = a + bi and w = c + di are two complex numbers, and suppose that $\overrightarrow{OP}$ is the vector corresponding to z and $\overrightarrow{OQ}$ is the vector corresponding to w (see figure below). Let R be the point which completes the

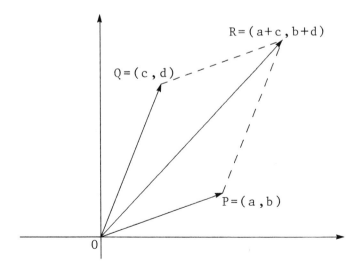

173

parallelogram having OP and OQ as adjacent sides. The x-coordinate of R is
a + c and the y-coordinate of R is b + d. That is, the vector $\overrightarrow{OR}$ corresponds
to the complex number (a + c) + (b + d)i and this number is equal to the sum
of z and w. Therefore, we have a geometrical interpretation for the sum of two
complex numbers: namely, it is a diagonal of a parallelogram (in the manner
just described). For example, if z and w are the complex numbers shown geometrically in the left figure below, their sum, z + w, corresponds to the vector
shown in the right figure below.

When a complex number is written in the form a + bi, a,b real, it is said
to be in *rectangular form* because of the way it is represented geometrically in
the plane.

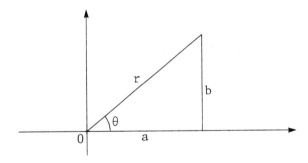

Let r denote the magnitude of a + bi. From the figure,

$$a = r \cos \theta, \quad \text{and} \quad b = r \sin \theta,$$

and it follows that

$$a + bi = r[\cos \theta + i \sin \theta].$$

174

This latter form for the representation of a complex number is called the *polar form* of a complex number because it explicitly displays the magnitude and the argument.

Example 2. Write each of the complex numbers below in polar form.

   a) $4 + 4i$

   b) $-\sqrt{7} + \sqrt{21}\, i$

Solution. For a), $r = \sqrt{(4)^2 + (4)^2} = 4\sqrt{2}$, and

$$4 + 4i = 4\sqrt{2}\, \left[\frac{4}{4\sqrt{2}} + \frac{4}{4\sqrt{2}}\, i\, \right].$$

It follows that $\cos \theta = 1/\sqrt{2}$, $\sin \theta = 1/\sqrt{2}$, and therefore $\theta = \pi/4$. Thus,

$$4 + 4i = 4\sqrt{2}\, [\cos \pi/4 + i \sin \pi/4].$$

For b), $r = \sqrt{(-\sqrt{7})^2 + (\sqrt{21})^2} = \sqrt{7 + 21} = \sqrt{28} = 2\sqrt{7}$, so

$$-\sqrt{7} + \sqrt{21}\, i = 2\sqrt{7}\, [-\sqrt{7}/2\sqrt{7} + (\sqrt{21}/2\sqrt{7})i]$$

$$= 2\sqrt{7}\, [-1/2 + (\sqrt{3}/2)i].$$

In this case $\cos \theta = -1/2$ and $\sin \theta = \sqrt{3}/2$, so that $\theta = 2\pi/3$, and we have

$$-\sqrt{7} + \sqrt{21}\, i = 2\sqrt{7}\, [\cos 2\pi/3 + i \sin 2\pi/3].$$

Suppose that $z = r[\cos \theta + i \sin \theta]$ and $w = s[\cos \phi + i \sin \phi]$ are two arbitrary complex numbers. Then

$$zw = rs[\cos \theta + i \sin \theta][\cos \phi + i \sin \phi]$$

$$= rs[(\cos \theta \cos \phi - \sin \theta \sin \phi) + i(\sin \theta \cos \phi + \cos \theta \sin \phi)]$$

$$= rs[\cos(\theta + \phi) + i \sin(\theta + \phi)].$$

This computation shows that the magnitude of the product of two complex numbers is equal to the product of the magnitudes of the two numbers, and the argument of the product is the sum of the arguments of the two factors.

More generally, if $z_1, \ldots, z_n$ are complex numbers, the magnitude of their product $z_1 z_2 \ldots z_n$ is the product of the magnitudes of each of the factors, and the argument of the product $z_1 \ldots z_n$ is the sum of the arguments of the factors.

Thus, in Example 2, the magnitude of

$$(4 + 4i)(-\sqrt{7} + \sqrt{21}\ i)$$

is equal to $(4\sqrt{2})(2\sqrt{7}) = 8\sqrt{14}$, and the argument is $\pi/4 + 2\pi/3 = 11\pi/12$ (see figure).

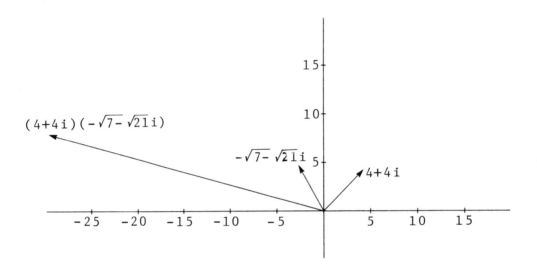

Example 3. Show that for $k = 1, 2, 3, 4,$ and 5

$$z_k = \cos\left(\frac{2k\pi}{5}\right) + i\ \sin\left(\frac{2k\pi}{5}\right)$$

satisfies the equation

$$z^5 = 1.$$

Solution. z_k is a complex number whose magnitude is $\sqrt{\cos^2\left(\frac{2k\pi}{5}\right) + \sin^2\left(\frac{2k\pi}{5}\right)}$ $= \sqrt{1} = 1$, and whose argument is $2k\pi/5$. The magnitude of $z_k{}^5$ is therefore equal to 1 (since the magnitude of each factor is 1) and the argument of $z_k{}^5$ is five times the argument of z_k: $5(2k\pi/5) = 2k\pi$. Thus, for each k,

$z_k^5 = \cos(2k\pi) + i \sin(2k\pi) = 1.$

The equation $z^5 = 1$, or equivalently, $z^5 - 1 = 0$, has five roots, the fifth roots of unity, and these are z_1, z_2, z_3, z_4, and z_5 as defined above. The figure below shows that they are uniformly distributed about the unit circle with $z_5 = 1$.

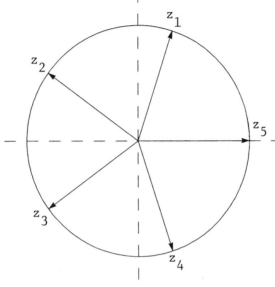

EXERCISES

Perform the indicated operations and write the result in rectangular form.

1. a) $(3 + 2i) - (4 + 3i)$

    b) $(3 + 2i)(4 + 3i)$

    c) $\dfrac{3 + 2i}{4 + 3i}$

2. a) $3(4 + i) + 4(2 - i)$

    b) $(4 + i)(2 - i)$

    c) $\dfrac{4 + i}{2 - i}$

3. a) $\dfrac{4 - 2i}{3 + 4i} - \dfrac{2 - i}{2 + i}$

    b) $i^5 + (2i)^3 + 4i$

    c) $\left(\dfrac{2i}{1 + i}\right)^3$

4. Find the magnitude and argument of each of the following complex numbers and then write each of the numbers in polar form.

    a) $-\sqrt{3} + i$ b) $4i$ c) $2 - 2i$.

5. Draw the vectors which correspond to the complex numbers $4 - 3i$ and $3 + i$

177

and compute their sum geometrically (that is, by drawing the appropriate parallelogram).

6. Same as Exercise 5 for $2 + 5i$ and $-4 + 2i$.

7. Draw the vectors which correspond to $\alpha = \sqrt{3} + i$ and $\beta = -\frac{3}{2} + \frac{3}{2}\sqrt{3}\,i$. Find the magnitude and argument of α and β, and draw the vector corresponding to $\alpha\beta$ (what is its magnitude and argument?). Verify this by direct computation.

8. Same as Exercise 7 for $\alpha = 1 + i$ and $\beta = -1 + i$.

9. Complex numbers with magnitude 1 are easy to multiply because geometrically we need only add the arguments. Draw the vector which represents $z = \frac{\sqrt{2}}{2} + \frac{\sqrt{2}}{2}\,i$ and compute the quantities below by using the geometrical interpretation of product.

a) z^2

b) z^3

c) z^4

d) $\frac{1}{z}$ (Where should the complex number $1/z$ be so that $(\frac{1}{z})z = 1$?)

e) z^8

10. In a manner analogous to Example 3, find all six roots of $z^6 = 1$. Represent them geometrically as vectors.

ANSWERS

1. a) $-1 - i$, b) $6 + 17i$, c) $18/25 - (1/25)i$ 2. a) $20 - i$, b) $9 - 2i$,

c) $7/5 + (6/5)i$ 3. a) $-11/25 - (2/25)i$, b) $-3i$, c) $-2 + 2i$

4. a) $2[\cos 5\pi/6 + i \sin 5\pi/6]$, b) $4[\cos \pi/2 + i \sin \pi/2]$,

c) $2\sqrt{2}[\cos 7\pi/4 + i \sin 7\pi/4]$

5.

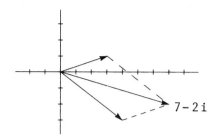

7 - 2i

6.

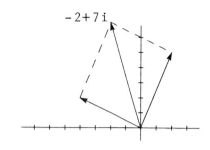

-2 + 7i

7.

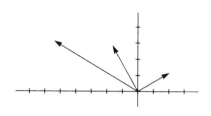

$|\alpha| = 2, \quad |\beta| = 3$

$\arg \alpha = \pi/6, \quad \arg \beta = 2\pi/3$

$|\alpha\beta| = 6$

$\arg(\alpha\beta) = 5\pi/6$

$\alpha\beta = -3\sqrt{3} + 3i$

8.

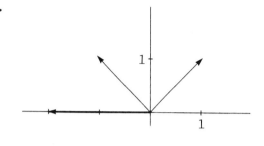

$|\alpha| = \sqrt{2}, \quad |\beta| = \sqrt{2}$

$\arg \alpha = \pi/4, \quad \arg \beta = 3\pi/4$

$|\alpha\beta| = 2$

$\arg(\alpha\beta) = \pi$

$\alpha\beta = -2$

9.

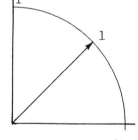

a) $z^2 = i$, b) $z^3 = -\sqrt{2}/2 + (\sqrt{2}/2)i$,

c) $z^4 = -1$, d) $1/z = \sqrt{2}/2 - (\sqrt{2}/2)i$

e) $z^8 = 1$.

10. Let $z_k = \cos 2k\pi/6 + i \sin 2k\pi/6$, for k = 1,2,3,4,5,6. Then $|z_k| = 1$, and argument $z_k = 2k\pi/6$. Therefore z_k^6 has argument $6(2k\pi/6) = 2k\pi$, so that $z_6^6 = 1[\cos 2k\pi + i \sin 2k\pi] = 1$.

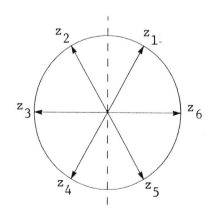

REFERENCES

If you need more explanation and practice on various sections of this
REFRESHER, you should refer to an algebra, trigonometry, or precalculus text-
book, such as those listed below.

Flanders, Harley, and Justine J. Price, *Algebra and Trigonometry*, Academic
Press, 1975.
> A beautiful book with informal presentations that are un-
> cluttered, easy to read, and to the point. The emphasis
> is on practical skills and problem solving.

Keedy, Mervin L., and Marvin L. Bittinger, *Fundamental Algebra and Trigonometry*,
Addison-Wesley, 1977.
> A pleasing format with an unusually large collection of
> examples and exercises.

Lial, Margaret L., and Charles D. Miller, *Algebra and Trigonometry*, Scott,
Foresman and Company, 1978.
> A very nice book, a textbook textbook, with thorough ex-
> planations given in a thoughtful and careful manner.

Munem, M. A., and J. P. Yizze, *Precalculus, Third Edition*, Worth Publishers,
1978.
> A somewhat formal approach organized within sections by
> definitions, theorems, proofs, examples, and exercises.

Peluso, Ada, *Background for Calculus*, Kendall/Hunt, 1978.
> An excellent reference for REFRESHER readers. Like the
> REFRESHER, the examples and exercises come from the con-
> text of calculus; however, the presentations require no
> previous acquaintance with algebra or trigonometry.

Swokowski, Earl, *Precalculus Mathematics, A Functional Approach*, Prindle,
Weber & Schmidt, 1973.
> A solid, academic presentation; somewhat dated by the lack
> of exercises requiring the hand calculator.

POST-TEST: Over Part I

(Answers on page 188)

1. Perform the operations: $2 - 3(4 - 5(3 - 5))$.

2. Simplify: $\dfrac{\dfrac{1}{ab} - \dfrac{1}{ac}}{\dfrac{1}{bc} - \dfrac{1}{ac}}$.

3. Solve for x: $\dfrac{360}{2\pi} = \dfrac{50}{x}$.

4. Solve for T: $\dfrac{T - S}{T} = \dfrac{a}{b}$.

5. Sketch the graph of $y = 4 - x^2$.

6. Write the equation of the line through the points $(5,-2)$ and $(-2,-3)$, and put the resulting equation in slope-intercept form ($y = mx + b$).

7. Find all values of x that satisfy each of the following statements:

   (i) $|3x - 1| = 2$ 　　　　　　　　(ii) $\dfrac{3}{|x - 1|} < 1$

8. Find all values for a for which the distance from $(a,2)$ to $(4,a)$ is greater than the distance from $(a,5)$ to $(7,a)$.

POST-TEST: Over Part II

(Answers on page 189)

1. Express $\left(\dfrac{2x^{-2}}{-y}\right)^{-3} \left(\dfrac{3y}{-x^{-1}}\right)^{2}$ as a simple fraction with no negative exponents.

2. Add and simplify: $x^{-1} + (1 + x)^{-1}$.

3. Estimate $\dfrac{(30,000)^2(.03)^3}{(400)(.0002)}$ and express your answer in scientific notation.

4. Simplify: $\dfrac{4x(x + 1)^3(x^2 + 2) - 3(x + 1)^2(x^2 + 2)^2}{(x + 1)^6}$.

5. Simplify: $\dfrac{\sqrt{3x^2y}\ \sqrt{10xy^4}}{\sqrt{8x^3y^3}}$.

6. Express $\sqrt[6]{x^2\sqrt{x}}$ as a rational power of x.

7. If $f(x) = (1 + x)^{-2/3}$, compute $f(26) - f(7)$.

8. If $f(x) = \dfrac{2x + 1}{x + 1}$, compute $\dfrac{f(x + h) - f(x)}{h}$ and simplify.

9. Solve for y': $(2x + y)(1 + 2y') - (x + 2y)(2 + y') = 0$.

POST-TEST: Over Part III

(Answers on page 189)

1. Solve for x: $x + \sqrt{x} - 2 = 0$. (Let $y = \sqrt{x}$.)

2. Solve for x: $x^3 + 2x^2 - x - 2 = 0$.

3. Solve the inequality $x(x - 1) \leq x + 8$.

4. Find where the curves $y = \frac{1}{2}x^2$ and $y = x + 4$ intersect. Draw a sketch of the area bounded by these two curves.

5. Which is longer: the length or the circumference of a can of tennis balls? (A tennis ball can is cylindrical and tightly holds a column of three tennis balls.)

6. The air resistance F of a moving object is approximately proportional to the square of the speed v of the object. Express this algebraically. Compare the air resistance of a car traveling 40 miles per hour with that of a car traveling 60 miles per hour.

7. A page is to contain 30 square inches of print. The margins at the top and bottom are 1 1/2 inches; the sides, 1 inch. Write the area of the page as a function of the width of the printed material.

8. Express $3x^2 - x$ in the form $A(x - h)^2 + k$.

9. Sketch the graph of $2x^2 + 3y^2 = 5$.

POST-TEST: Over Part IV

(Answers on page 189)

1. Change $\log_a b = c$ to an equivalent exponential equation.

2. Express $\log_a \dfrac{x^2 y}{\sqrt{yz}}$ in terms of logarithms of x, y, and z.

3. Solve for x:

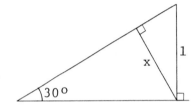

4. Express $[\cos \dfrac{5\pi}{6} - \sin \dfrac{5\pi}{6}] - [\cos \dfrac{\pi}{2} - \sin \dfrac{\pi}{2}]$ in terms of radicals.

5. Which, if any, of the following equations are valid for all θ?

   (a) $\cos(\theta + 2\pi) = \cos \theta$ (b) $\tan(\theta + 2\pi) = \tan \theta$ (c) $\cos (\dfrac{\pi}{2} + \theta) = \sin \theta$

6. Sketch the graph of $y = - \cos \theta$.

7. Solve for θ: $2 \cos \theta - 1 = 0$.

8. Express $\dfrac{1}{\sec \theta - \tan \theta} - \dfrac{1}{\sec \theta + \tan \theta}$ in terms of sines and cosines,

   then add and simplify.

9. Which, if any, of the following equations are valid for all θ?

   (a) $\sin \theta = 2 \sin \theta \cos \theta$ (b) $\cos^2 \dfrac{\theta}{2} = \dfrac{1 + \cos \theta}{2}$

   (c) $\sin 3\theta = \sin \theta \cos 2\theta + \sin 2\theta \cos \theta$

10. If $f(\theta) = \text{Arctan } \theta$, evaluate $f(1) - f(0)$.

11. Write $\dfrac{2 + 3i}{3 - 4i}$ in the form $a + bi$, a and b real.

ANSWERS

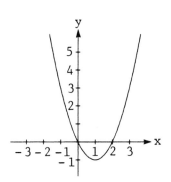
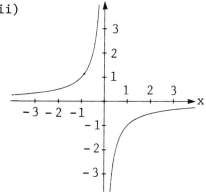

Table of Contents, Part III

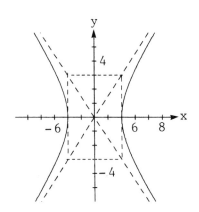

Table of Contents, Part IV

32. (i) 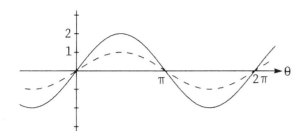   (ii)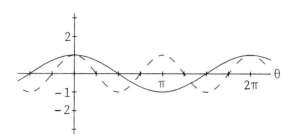

33. (i) $\theta = 2n\pi$, $n = 0, \pm1, \pm2, \ldots$ (ii) $\theta = \pi/12 + n\pi$, $5\pi/12 + n\pi$,

$n = 0, \pm1, \pm2, \ldots$ 34. (i) $2 \csc x$ (ii) $(4 + 3\sqrt{3})10$ 35. (i) $\sqrt{3}/2$

(ii) $\dfrac{x}{\sqrt{1 - x^2}}$ 36. (i) $1 - i$ (ii) $8; 120°$

Diagnostic Test, Part I

1. $11\sqrt{7} - 10\sqrt{3} + 36$ 2. $5a - 5b - 4$ 3. a/b 4. $x = 20$

5.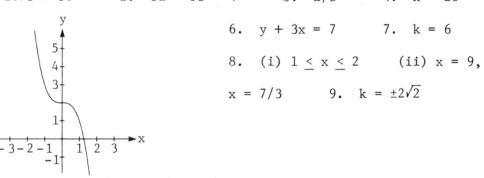

6. $y + 3x = 7$ 7. $k = 6$

8. (i) $1 \le x \le 2$ (ii) $x = 9$,

$x = 7/3$ 9. $k = \pm2\sqrt{2}$

Diagnostic Test, Part II

1. $\dfrac{-x^4}{4y^7}$ 2. $\dfrac{2x}{(x + 1)^3}$ 3. 5.87×10^{12} 4. $8xy - 3x^2 + 3y^2$

5. $\dfrac{x - 3}{2x + 1}$ 6. $\dfrac{x(x^2 + 8)}{\sqrt{x^2 + 4}}$ 7. $2xy \sqrt[3]{3x^2y^2}$ 8. $f(1 + x) = -\dfrac{1}{x}$,

$f(1 - x) = \dfrac{1}{x}$ 9. $x = \dfrac{1 - y}{y - 2}$

Diagnostic Test, Part III

1. $x = \dfrac{-1 \pm \sqrt{13}}{6}$ 2. $x = -4, 1, 3$ 3. $x \le -1$, $x \ge 1/2$

4. $A = -1/4$, $B = 5/4$ 5. $V = 32\pi/3$ 6. (a) Less than 40 because more

time was spent going at the slower speed.   (b) $50 = d/T_1$, $30 = d/T_2$

(c) 37.5 miles per hour   7. $A = h^2/4$

8. $y = -2\left(x - \dfrac{5}{4}\right)^2 + \dfrac{17}{8}$

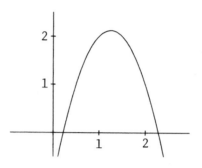

Diagnostic Test, Part IV

1. $x = 8$ 2. $\log_a\left(\dfrac{x^2\sqrt{x+2}}{(x-1)^3}\right)$ 3. $2(\sqrt{3} - 1)$ 4. $\sqrt{2} + 1/\sqrt{3}$

5. (b)   6.

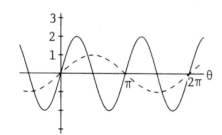

7. $\theta = \dfrac{(2n+1)\pi}{6}$,

$n = 1, 2, 3, 4, 5, 6$

8. $\sec\theta$ 9. (a)

10. $\sqrt{1 - x^2}$ 11. $4, 240°$

Post-Test, Part I

1. -40 2. $\dfrac{c - b}{a - b}$ 3. $x = 5\pi/18$ 4. $T = \dfrac{Sb}{b - a}$

5.

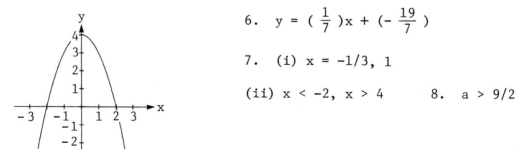

6. $y = \left(\dfrac{1}{7}\right)x + \left(-\dfrac{19}{7}\right)$

7. (i) $x = -1/3$, 1

(ii) $x < -2$, $x > 4$ 8. $a > 9/2$

Post-Test, Part II

1. $-\frac{9}{8} x^8 y^5$

2. $\frac{1 + 2x}{x(1 + x)}$

3. 3×10^5

4. $\frac{(x^2 + 2)(x^2 + 4x - 6)}{(x + 1)^4}$

5. $(\frac{\sqrt{15}}{2})y$

6. $x^{5/12}$

7. $-5/36$

8. $\frac{1}{(x + 1)(x + h + 1)}$

9. $y' = y/x$

Post-Test, Part III

1. $x = 1$

2. $x = -2, -1, 1$

3. $-2 \le x \le 4$

4. $x = -2, y = 2$ and $x = 4, y = 8$

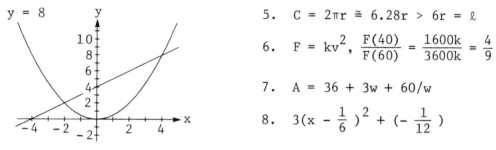

5. $C = 2\pi r \cong 6.28r > 6r = \ell$

6. $F = kv^2$, $\frac{F(40)}{F(60)} = \frac{1600k}{3600k} = \frac{4}{9}$

7. $A = 36 + 3w + 60/w$

8. $3(x - \frac{1}{6})^2 + (-\frac{1}{12})$

9. $\frac{x^2}{a^2} + \frac{y^2}{b^2} = 1$, $a = \sqrt{5/2} \cong 1.58$, $b = \sqrt{5/3} \cong 1.29$

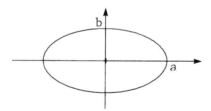

Post-Test, Part IV

1. $a^c = b$

2. $2 \log_a x + \frac{1}{2} \log_a y - \frac{1}{2} \log_a z$

3. $x = \sqrt{3}/2$

4. $(1 - \sqrt{3})/2$

5. (a), (b)

6.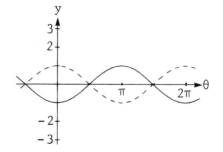

7. $\theta = \pi/3 + 2n\pi$, $\theta = -\pi/3 + 2n\pi$, $n = 0, \pm 1, \pm 2, \ldots$

8. $2 \tan \theta$

9. (b), (c)

10. $\pi/4$

11. $-\frac{6}{25} + \frac{17}{25} i$

INDEX